Practical Organic Chemistry
Volume-II

Practical Organic Chemistry
Volume-II

Sharda Pasricha
Associate Professor
Department of Chemistry
Sri Venkateswara College
University of Delhi
Delhi

Ankita Chaudhary
Assistant Professor
Department of Chemistry
Maitreyi College
University of Delhi
Delhi

I.K. International Pvt. Ltd.
NEW DELHI

Published by
I.K. International Pvt. Ltd.
4435-36/7, Ansari Road, Daryaganj
New Delhi-110 002 (India)
E-mail: info@ikinternational.com
Website: www.ikbooks.com

ISBN 978-93-90620-22-7

Published/Printed by Krishan Makhijani for I.K. International Pvt. Ltd., 4435-36/7, Ansari Road, Daryaganj, New Delhi–110 002.

This book is dedicated to my mother,
Mrs. Raj Jaspal and my husband, **Mr. Neeraj Pasricha**
who have constantly encouraged me and supported me
during the writing of this book.
Sharda Pasricha

Dedicated to my nieces:
Yuvika & Aditi for making my days brighter
and bringing joy to my heart!
Ankita Chaudhary

Preface

In an organic chemistry laboratory, at undergraduate level, students are required to do isolation, separation, purification, synthesis, estimation and identification of organic compounds. The basic techniques of separation, purification and synthesis have been discussed at length in the first volume of this book. This second volume focuses on isolation, quantitative estimation and identification of unknown organic compounds. This along with the topics handled in the first volume will equip the students with all the necessary acumen required to handle challenges in the undergraduate, postgraduate as well as the research labs. Like the first, this volume is written keeping in mind the latest UGC syllabus and the previous CBCS syllabus running in all the central universities across India as well as other universities like BHU, PU, KU, MDU, etc.

In compiling this book, we have consulted all types of sources and references available to us, including numerous practical organic chemistry textbooks, research papers available in the journals, MSDS of the chemical compounds and practical manuals of the educational institutions, etc. The authors are indebted to all these sources including ones where it was not possible to give individual acknowledgements. The students are advised to further consult the original references as some modifications to original work may have to be done to suite the laboratory requirements of an undergraduate class.

The present volume has been divided into four chapters. Chapter 1 is devoted to quantitative estimation of some selected organic compounds through the medium of their functional groups. Chapter 2 is largely concerned with the methods of extraction and isolation of some useful organic compounds and biomolecules from their natural sources. These experiments expose the undergraduate students to microscale extraction techniques. Chapter 3 deals with biochemical estimations involving amino acids, proteins, enzyme action and factors affecting enzyme action. Chapter 4 discusses methods for qualitative organic analysis. Identification of unknown organic compound forms an important part of the organic chemistry lab at the undergraduate level. With the advent of chromatography and modern spectroscopic techniques, practising chemists now have a shorter and a better route to identify the unknown organic compounds. However, the traditional chemical methods are still popular at the undergraduate level, as they provide the necessary training to approach a problem in a logical and systematic manner. This book discusses some of the most common chemical reactions for qualitative analysis of organic compounds based on the nature of the functional group present in them. A systematic scheme and a detailed description of the steps involved in establishing the structure of an unknown compound through qualitative analysis has been discussed. A brief idea of how IR and NMR spectroscopic data can be used to augment the knowledge obtained from chemical methods is given at the end of each subsection covering a specific functional group. In order to obviate the drawbacks of traditional and hazardous chemical reactions, green chemistry approach of using spot tests and greener alternatives have been discussed in this chapter wherever possible. Disposal strategies, wherever available, have been given after every chemical test. A special note of the things to remember, while performing experiments, based on the experience of the authors in undergraduate classes, has been included especially for the benefit of the students. These features distinguish this text from several other texts written for the same target audience. Appendix I gives the classified tables for the organic compounds. Appendix II describes the methods of preparation of various reagents used in an undergraduate organic chemistry lab. Appendix III gives the preparation of some common buffers used in biochemical estimations.

No claim is made that this volume deals with methods for estimations, extraction and qualitative analysis of all types of organic compounds. However, this basic idea will help students to improvise and approach a wider range of compounds in a logical and systematic manner. All suggestions, errors, improvisations and new methodologies may be sent via email to: shardap@gmail.com.

We are thankful to all our organic chemistry teachers especially, Prof. S K Grover, Prof. J M Khurana, Dr. V. Krishnamurthy for all the knowledge they have shared with us. Their teachings have inspired us to write this text in two volumes.

Sharda Pasricha
Ankita Chaudhary

Contents

4. Qualitative Analysis   86

Quantitative Estimation

1.1 INTRODUCTION

This section deals with quantitative analysis of certain types of organic compounds normally estimated in an undergraduate laboratory. The functional groups present in them serve as medium of these estimations.

1.2 ESTIMATION OF AMINES

Amines can be estimated in the laboratory by any one of the following methods:

(a) Acetylation

(b) Bromination (for aromatic amines)

1.2.1 Estimation of the Number of Amino Groups Present in the Given Sample by Acetylation Method

Basic Lab Operations: Refluxing and acid-base titration.

Estimated Time: 90 minutes

Chemicals Required: Aniline, acetic anhydride, glacial acetic acid, oxalic acid, sodium hydroxide, phenolphthalein.

What You Should Know: Acetylation of primary as well as secondary amines, phenols and alcohols involves replacement of the active hydrogen by acetyl group ($-CH_3CO$). This method offers a means of estimation based on the number of equivalents of acetic anhydride used per mole of the amine/phenol/alcohol. Tertiary amines cannot be estimated by this method due to lack of replaceable hydrogen.

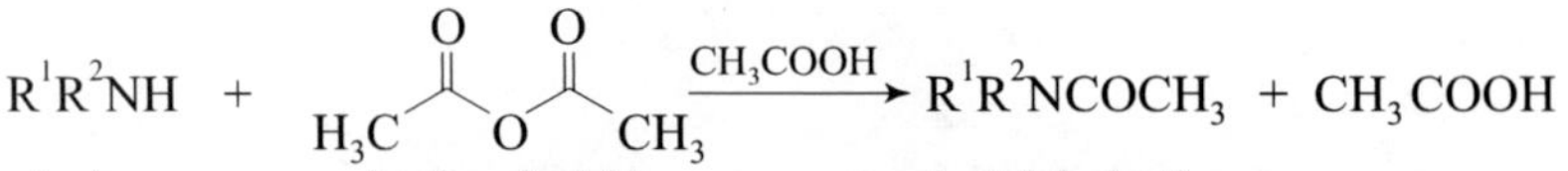

$$R^1R^2NH \ + \ \underset{\text{Acetic anhydride}}{H_3C-\underset{O}{\overset{O}{\parallel}}-O-\underset{O}{\overset{O}{\parallel}}-CH_3} \ \xrightarrow{CH_3COOH} \ \underset{\text{Acetyl derivative}}{R^1R^2NCOCH_3} \ + \ CH_3COOH$$

Amine

$R^1, R^2 = H$, Alkyl, Aryl

A known weight of the substance (**test sample**) is treated with excess but an exact amount of freshly prepared acetylating mixture (two volumes of freshly distilled acetic anhydride and one volume of glacial acetic acid). The active hydrogen in the molecule is replaced by acetyl group. On completion of reaction, the unreacted acetic anhydride is hydrolyzed to acetic acid by water.

$$\underset{\text{Acetic anhydride}}{(CH_3CO)_2O} \ + \ H_2O \ \longrightarrow \ \underset{\text{Acetic acid}}{2\ CH_3COOH}$$

The acetic acid produced by hydrolysis is estimated volumetrically with a standardized sodium hydroxide solution. A simultaneous **blank experiment** (without amine/phenol/alcohol) is performed to know the total amount of acetic acid formed from the sample of acetic anhydride used in the experiment. The difference in the titre values of sodium hydroxide used equals the amount of acetic acid actually used for the acetylation. If the molecular mass of the test compound is known, number of alcoholic/phenolic/amino groups can be calculated or strength of the solution can be determined. Some diacetyl derivative may also be formed in case of primary amines. However, this is of no consequence because on refluxing with hot water, it is converted to monoacetyl derivative.

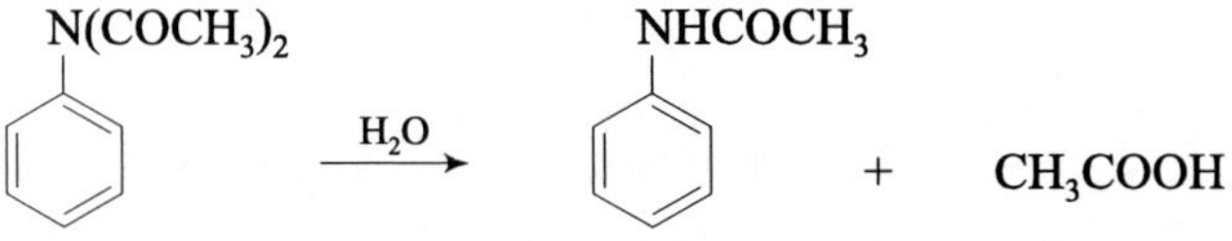

$$\underset{\substack{\text{N-Acetyl-N-phenylacetamide}\\\text{(Diacetyl derivative)}}}{\text{N(COCH}_3)_2\text{—C}_6\text{H}_5} \ \xrightarrow{H_2O} \ \underset{\text{Acetanilide}}{\text{NHCOCH}_3\text{—C}_6\text{H}_5} \ + \ \underset{\text{Acetic acid}}{CH_3COOH}$$

Role of blank titration

- Absolute concentration of acetic anhydride in solution need not be determined as difference of titre values of NaOH for blank and test samples gives the actual amount of acetic anhydride, in the form of acetic acid, used for acetylation.

- Indeterminate loss of reagents due to chemical reaction with glass material, slight absorption by corks will be identical in both test and blank samples. Therefore, these do not contribute to any additional losses.

Role of acetic acid

- Acts as catalyst in the reaction (see Volume 1, *section 6.7,* Acetylation)

Chemical Reaction Involved

$$\underset{\text{Aniline}}{NH_2\text{-}C_6H_5} + \underset{\text{Acetic anhydride}}{(CH_3CO)_2O} \xrightarrow{CH_3COOH} \underset{\text{Acetanilide}}{C_6H_5\text{-}NHCOCH_3} + \underset{\text{Acetic acid}}{CH_3COOH}$$

Experimental Procedure: Prepare a standard solution of oxalic acid (100 mL, N/2) and sodium hydroxide solution (~1.0 N, 200 mL). Take the sodium hydroxide solution in the burette and pipette out standard oxalic acid solution (10.0 mL) in the conical flask. Titrate it against sodium hydroxide solution using phenolphthalein as the indicator. Determine the normality of sodium hydroxide solution.

Take a dry boiling tube and prepare the acetylating mixture by mixing 2 volumes of freshly distilled acetic anhydride with one volume of glacial acetic acid. Shake the contents vigorously and keep the reagent aside. Subsequently, take two dry round bottomed flasks (R.B. flask), **A** and **B**. Weigh **A** and run down aniline (1.5 mL) in it and weigh the R.B. flask again. Note the mass of aniline transferred. Cool flask **A** containing aniline in ice and add dropwise acetylating mixture (4.0 mL), with shaking, to flask **A** and add same amount of acetylating mixture to flask **B**. Put 1-2 pumice stones in each of the two flasks and connect them to reflux water condenser. Reflux both the mixtures on boiling water bath for 30 minutes, with occasional shaking. After 30 minutes, cool both the flasks slightly and dilute each with hot water (50.0 mL). Reflux the contents of both the flasks for another 10 minutes. Cool, wash the bottom portion of the condenser with distilled water and allow the washing to run down into the respective R.B. flasks. Now, titrate **A** with 1N standardised NaOH using phenolphthalein as indicator. Shake the contents of the flask vigorously during titration in order to keep crystals of acetanilide dispersed in aqueous solution. Note the volume of NaOH used. Repeat the same procedure with the blank sample (**B**) and again note the volume of NaOH consumed.

Observations and Calculations

Mass of the aniline dissolved $= w$ g

Normality of the NaOH solution $= N$

Volume of NaOH used for blank sample $= V$ mL

Volume of NaOH used with test sample $= V_1$ mL

$V - V_1$ mL of N normal NaOH = Volume of acetic acid used for acetylation of w g of aniline

1000.0 mL of $1N$ NaOH = 1.0 g mole of NaOH = 1.0 g mole of acetic acid = 1 NH_2 group

$$\therefore \quad V - V_1 \text{ mL of } N \text{ normal NaOH} = \frac{(V - V_1) \times N}{1000} \text{ NH}_2 \text{ groups} = w\,\text{g}$$

$$\therefore \quad w\,\text{g of aniline} = \frac{(V - V_1) \times N}{1000} \text{ NH}_2 \text{ groups}$$

$$1.0 \text{ g mole of aniline } (M) = \frac{(V - V_1) \times N \times M}{1000 \times w} \text{ NH}_2 \text{ groups}$$

where, $\quad M = \text{Molar mass of the aniline.}$

Calculation of Molar mass of aniline

$$V - V_1 \text{ mL of } N \text{ normal NaOH} = w\,\text{g of aniline}$$

$$\therefore \quad 1000.0 \text{ mL of } 1.0 \, N \text{ NaOH} = \frac{w \times 1000}{(V - V_1) \times N} \, g \text{ of aniline}$$

$$M = \frac{w \times 1000}{(V - V_1) \times N} \, g \text{ of aniline}$$

Note

- Excess of acetic anhydride used ensures that the reaction goes to completion. If acetic anhydride is used alone, then the results are higher in case of primary amines due to formation of diacetyl derivative.

- Aniline should be cooled in ice before addition of the acetylating agent otherwise the heat generated in the reaction will lead to spurting out of acetylating mixture.

- For better results, dilute the mixture obtained after acetylation to 100 mL with distilled water, in a standard flask, and take 10.0 mL of the aliquot for the titration with sodium hydroxide. Repeat the procedure to get three concordant values. This avoids use of excessive amounts of alkali and minimizes experimentation errors.

- Hot water is added, after refluxing for 30 minutes, to prevent acetanilide from precipitating as hard crystals, which interfere with the titration.

- The crystals of acetanilide may interfere with the titration. Therefore, it is important to shake the contents vigorously during the titration to ensure that all the unreacted acetic acid is neutralized by NaOH.

- Reagent amounts should be doubled if amine contains two amino groups.

- Use of reflux condenser is not essential for acetylation as not much of evaporation takes place on the water bath, under these conditions. However, use of water condenser prevents condensing of steam.

- Acetic anhydride-dry pyridine is a useful reagent for estimation of amines, phenols and alcohols. However, its use is avoided here due to high cost of pyridine.

Result: The given amine contains .... amino groups.

Utility of This Exercise

- Estimation of commercial samples of amines, phenols and alcohols.

Lessons to Take Home

The students will be able to do the following after doing this experiment:

- Understand application of acetylation in amines, alcohols and phenols.
- Explain the need for blank titration in such quantitative exercises.
- Determine percentage purity of a given sample of amine, alcohol and phenol.
- Predict the number of free amino/hydroxyl groups present in a given sample, provided the molar mass of the sample is known.

1.2.2 Estimation of the Amount of Aniline Present in the Given Sample by Bromate-Bromide Method

Basic Lab Operations: Iodometric titration.

Estimated Time: 90 minutes

Chemicals Required: Potassium bromate, potassium bromide, potassium iodide, sodium thiosulphate, potassium dichromate, aniline, concentrated hydrochloric acid and sulphuric acid.

What You Should Know: Aromatic amines or phenols are highly activated for an electrophilic substitution reaction with bromine, due to the presence of strong electron donating amino/phenolic group. Such compounds undergo bromination reaction at *o*- and *p*-positions, when treated with excess of bromine. Polar solvents like water and acetic acid further facilitate the reaction by generating bromonium ion, which is the electrophile in such reactions.

Bromine is a volatile liquid and hence keeps evaporating. It is therefore advisable to generate bromine *in situ*, in order to minimize the error due to loss of bromine. Potassium bromide liberates bromine *in situ* on treatment with oxidizing agent, like potassium bromate, in the presence of hydrochloric acid. The liberated bromine reacts with the activated *o*- and *p*-positions of aniline or phenol to form 2,4,6 tribromoaniline or 2,4,6 tribromophenol. Excess of unreacted bromine is converted to iodine on reaction with potassium iodide. The liberated iodine, which equals the unreacted bromine (V_1 mL), is estimated iodometrically by titration with standardized sodium thiosulphate solution.

A blank titration is simultaneously performed to know the total amount of bromine generated from the added volumes of potassium bromate and potassium bromide (V mL). The difference of the two titre values ($V - V_1$ mL) equals the total volume of bromine which has reacted with aniline or phenol.

This volume can be used to calculate the strength of the given solution of phenol or amine.

Chemical Reaction Involved

$$5KBr + KBrO_3 + 6HCl \longrightarrow 3Br_2 + 6KCl + 3H_2O$$

Aniline $+ 3Br_2 \longrightarrow$ 2,4,6-Tribromoaniline $+ 3HBr$

$$Br_2 + 2KI \longrightarrow I_2 + 2KBr$$
Unreacted

$$I_2 + 2Na_2S_2O_3 \longrightarrow 2NaI + Na_2S_4O_6$$

Experimental Procedure: Prepare a standard solution of potassium dichromate (100 mL, N/10), sodium thiosulphate (100 mL, ~N/10) and potassium iodide (50 mL, 5% and 10% each), aniline solution (1.0 g/L) and bromate -bromide solution (100 mL, 0.2 N).

Take a clean standard flask (1.0 L) and transfer freshly distilled aniline (1.0 g) to it. Dissolve aniline in minimum volume of dilute hydrochloric acid (prepared by adding AR HCl to distilled water taken in a beaker). Dilute with distilled water up to the mark and replace the stopper (Do not forget to homogenize the solution!). Keep the solution aside.

Take the sodium thiosulphate solution in the burette. Pipette out standard potassium dichromate solution (10.0 mL) in the conical flask, add sulphuric acid (1.0 N, 10.0 mL), potassium iodide solution (10.0 mL, 5%). Cover the conical flask and keep it undisturbed for 10 minutes. The solution becomes dark yellow due to liberated iodine. Titrate the liberated iodine against sodium thiosulphate solution till solution becomes straw yellow. Now, add 5-6 drops of starch and continue addition of sodium thiosulphate solution till the solution becomes light green. Repeat to get three concordant values. Determine the normality of sodium thiosulphate solution.

Now take three clean iodine flasks. In two of those iodine flasks (**test samples**), add aniline (10.0 mL), followed by bromate-bromide solution (10.0 mL) and then distilled water (10.0 mL). Mix the contents and then add concentrated HCl (2.0 mL) to both the flasks and immediately stopper them. In the third flask, prepare a blank sample by taking bromate-bromide solution (10.0 mL), distilled water water (20.0 mL) and concentrated hydrochloric acid (2.0 mL) (do not add aniline soluton!). Immediately stopper the flask and allow all the three flasks to stand for 25 minutes in water bath at room temperature. Shake the contents occasionally. After 25 minutes, add potassium iodide solution (5.0 mL, 10%) to all the three flasks and immediately place the stopper again. Allow the contents to stand for another 10 minutes. A white precipitate of 2,4,6-tribromoaniline will settle down in the two test samples and the flasks acquire a distinct yellow-orange colour due to liberation of iodine.Rinse the stopper and neck of the flask with water (5.0 mL) and allow all the washings to run into the flask. Titrate the test samples one by one with standardized sodium thiosulphate solution till straw yellow coloured solution is obtained. Add starch solution (5-6 drops, immediately observe blue-black colour) and continue the addition of sodium thiosulphate till the solution becomes colourless. Note the volume of thiosulphate solution used (V_x and V_y). Take average of the two values as V_1. Also repeat the same procedure with the blank sample and again note the volume of sodium thiosulphate used (V mL).

Observations and Calculations

Normality of the standardized sodium thiosulphate solution = N

Volume of sodium thiosulphate solution used for blank titration = V mL

Volume of sodium thiosulphate solution used with test sample = V_1 mL

$V - V_1$ mL of 1.0 N sodium thiosulphate solution = Total amount of bromine reacted

$$1000.0 \text{ mL of } 1.0 \ N \text{ sodium thiosulphate} = 1.0 \text{ g equivalent of iodine}$$
$$= 1.0 \text{ g equivalent of bromine}$$
$$= 1/6 \text{ moles of aniline}$$

$$V - V_1 \text{ mL of } N \text{ normal Na}_2\text{S}_2\text{O}_3 = (V - V_1) \times \frac{N}{1000 \times 6} \text{ moles of aniline}$$

$$= \frac{(V - V_1) \times N \times M}{6000} \text{ g of aniline}$$

where, M = molar mass of the aniline.

If, w = amount of aniline in 10.0 mL then,

$$w = \frac{(V - V_1) \times N \times M}{6 \times 1000} \text{ g of aniline}$$

$\therefore$ *Strength of aniline in the given solution* = $w \times 100.0$ *g/L*

% Purity of the given sample of aniline $= \dfrac{(V - V_1) \times N \times M}{6000 \times W} \times 100\%$

where, W = actual mass of aniline dissolved in 10.0 mL (Divide mass of aniline dissolved in one litre by 100 to get W)

Note

- For preparation of potassium bromate-potassium bromide reagent, See Appendix II.

- Large excess of potassium bromide ensures complete reduction of potassium bromate upon acidification. Also, it increases solvent power of the solution for free bromine.

- Shake the contents vigorously while carrying out the titration with test samples, this allows all the liberated iodine to react with sodium thiosulphate, without any interference from the tribromo derivative.

- Aniline should be transferred into the iodine flask with the help of burrete. (Do not pipette)

Result: The strength of the given solution of aniline was found to be ----------- g/L.

Utility of This Exercise

- Estimation of commercial samples of amines, phenols and alcohols.

Lessons to Take Home

- Understand the application of bromination in amines and phenols.

- Determine the percentage purity of the given sample of amine or phenol.

1.3   ESTIMATION OF PHENOLS

Like amines, phenols can also be estimated in the laboratory by any one of the following methods:

    (a) Acetylation

    (b) Bromination (for aromatic phenols)

1.3.1   Estimation of the Number of Phenolic Groups Present in a Given Sample by Acetylation Method

Basic Lab Operations: Refluxing and acid-base titration.

Estimated Time: 90 minutes

Chemicals Required: Phenol, acetic anhydride, dry pyridine, oxalic acid, sodium hydroxide, phenolphthalein.

What You Should Know: Like amines, acetylation of phenols can be carried out conveniently with acetic anhydride in the presence of dry pyridine as catalyst. The number of free –OH groups present in the molecule can be estimated from the number of equivalents of acetic anhydride used per mole of the phenol.

A known weight of the substance (**test sample**) is treated with excess but an exact amount of freshly prepared acetylating mixture (one volume of acetic anhydride and four volumes of dry pyridine). *N*-acetylpyridinium acetate formed by reaction of pyridine with acetic anhydride is the actual acetylating agent. The active hydrogen in the phenol is replaced by acetyl group. On completion of reaction, the unreacted *N*-acetylpyridinium acetate is hydrolysed by water to acetic acid and pyridinium acetate. The acetic acid produced is estimated volumetrically with a standardized sodium hydroxide solution. A simultaneous **blank experiment** (no phenol is added) is performed to know the total amount of acetic acid formed from the sample of acetic anhydride taken in the experiment. The difference in the titre values of sodium hydroxide used equals the amount of acetic acid actually used for the acetylation. If the molecular mass of the test compound is known, number of phenolic groups can be calculated or strength of the solution can be determined or else if the number of free –OH groups are known, the molecular mass of the unknown phenol can be calculated.

Role of Pyridine

- Acts as catalyst in the reaction by forming *N*-acetylpyridinium acetate, which is a better acetylating agent than acetic anhydride.
- It is used as a solvent in acetylation, being inert to the reagents.
- It removes the acid produced in the reaction by salt formation.

Chemical Reaction Involved

Experimental Procedure: Prepare a standard solution of oxalic acid (100 mL, N/2) and sodium hydroxide solution (~1N, 100 mL). Take the sodium hydroxide solution in the burette, pipette out oxalic acid solution (10.0 mL) in the conical flask. Titrate it against sodium hydroxide solution taking phenolphthalein as the indicator. Determine the normality of the sodium hydroxide solution.

Take a dry R.B. flask and prepare the acetylating mixture by mixing one volume of acetic anhydride with four volumes of dry of pyridine. Shake the contents vigorously and transfer the mixture to a clean and dry burette. Cover the mouth of the burette with sodalime guard tube.

Next, take two dry 100 mL round bottomed flasks, **A (test)** and **B (blank)**. Weigh **A** and add phenol (1.0 mL) or 2-naphthol (~ 1.0 g) to it and weigh the R.B. flask again. Note the mass of phenol transferred. Cool the flasks in ice and add the acetylating mixture (10.0 mL) dropwise with shaking to both **A** and **B**. Put 1-2 pumice stones in each R.B. flask and connect both the R.B. flasks to reflux water condenser. Reflux both the mixtures on the boiling water bath for 30 minutes, with occasional shaking. After 30 minutes, cool both the flasks and dilute each one with hot water (20.0 mL). A fine emulsion of phenyl acetate separates, in flask A, on dilution. Reflux the contents of both the flasks for another 10 minutes. Cool, wash the bottom portion of the condenser with distilled water and allow the washing to run down into the respective R.B. flasks. Now, titrate **A** with 1 N standardized NaOH solution using phenolphthalein as indicator. Shake the contents of the flask vigorously during titration to ensure that all the acetic acid formed on hydrolysis reacts will alkali. Note the volume of NaOH used. Repeat the same procedure with the blank sample (**B**) and again note the volume of NaOH consumed.

Observations and Calculations

Mass of the phenol dissolved = w g

Normality of the NaOH solution = N

Volume of NaOH used for blank sample = V mL

Volume of NaOH used with test sample = V_1 mL

$V - V_1$ mL of N normal NaOH = volume of acetic acid used for acetylation of w g of phenol

1000.0 mL of 1.0 N NaOH = 1.0 g mole of NaOH = 1.0 g mole of acetic acid = 1 hydroxyl group

$$\therefore \quad V - V_1 \text{ mL of } N \text{ normal NaOH} = \frac{(V - V_1) \times N}{1000} \text{ hydroxyl groups}$$

$$\therefore \quad w \text{ g of phenol} = \frac{(V - V_1) \times N}{1000} \text{ hydroxyl groups}$$

$$1 \text{ g mole of phenol } (M) = \frac{(V - V_1) \times N \times M}{1000 \times w} \text{ hydroxyl groups}$$

Calculation of Molar mass of Phenol

$V - V_1$ mL of N normal NaOH = w g of phenol

$\therefore$ 1000.0 mL of $1N$ NaOH = $\dfrac{w \times 1000}{(V - V_1) \times N}$ g of phenol

$$M = \dfrac{w \times 1000}{(V - V_1) \times N} \text{ g of phenol}$$

M = Molar mass of phenol

Note

- Dry pyridine over solid KOH and distill before use.
- For other points, see *section 1.2.1*

Result: The given phenol contains ---------- hydroxyl groups

Utility of This Exercise

See *section 1.2.1*

Lessons to Take Home

See *section 1.2.1*

1.3.2 Estimation of the Amount of Phenol Present in the Given Sample by Bromate-Bromide Method

Basic Lab Operations: Iodometric titration.

Estimated Time: 90 minutes

Chemicals Required: Potassium bromate, potassium bromide, potassium iodide, sodium thiosulphate, potassium dichromate, phenol, concentrated hydrochloric acid and sulphuric acid.

What You Should Know: The detailed theory for bromination of phenol is same as that for aniline, discussed in *section 1.2.2*, except that phenol reacts with potassium bromate-potassium bromide in the presence of concentrated hydrochloric acid to form 2,4,6-tribromophenol along with small amount of 2,4,6-tribromophenol bromide. This is however of no consequence because, when the reaction mixture is allowed to stand for some time in the acidic medium, in presence of potassium iodide, the bromide salt is converted to 2,4,6-tribromophenol. This rules out any error in estimating the amount of unreacted bromine.

Chemical Reaction Involved

$$5KBr + KBrO_3 + 6HCl \longrightarrow 3Br_2 + 6KCl + 3H_2O$$

OH $\quad$ + 3Br$_2$ $\longrightarrow$ Br—(OH)—Br + 3HBr

Phenol $\qquad$ 2,4,6-Tribromophenol

2,4,6-Tribromophenol bromide

$$Br_2 + 2KI \longrightarrow I_2 + 2KBr$$
Unreacted

$$I_2 + 2Na_2S_2O_3 \longrightarrow 2NaI + Na_2S_4O_6$$

Experimental Procedure: Prepare the reagents as detailed for estimation of amines, in *section 1.2.2*. Take a clean standard flask (1.0 L) and transfer phenol (1.0 g) into it. Dissolve the phenol in minimum volume of dilute sodium hydroxide solution (~20.0 mL, 10%). Dilute with distilled water up to the mark and replace the stopper. (Do not forget to homogenize the solution!) Keep the solution aside.

Take the sodium thiosulphate solution in the burette, pipette out standard potassium dichromate solution (10.0 mL, N/10) in a clean conical flask. Add sulphuric acid (1N, 10.0 mL), potassium iodide (10.0 mL, 5%). Cover the conical flask and keep it undisturbed for 10 minutes. The solution becomes dark yellow due to liberated iodine. Titrate the liberated iodine against sodium thiosulphate solution till solution becomes straw yellow. Add 5-6 drops of starch and continue addition of sodium thiosulphate solution till the solution turns light green. Repeat to get three concordant values. Determine the normality of the sodium thiosulphate solution.

Take three clean iodine flasks. In the first two iodine flasks (**test samples**), add phenol solution (10.0 mL), followed by bromate-bromide solution (10.0 mL) and then distilled water (10.0 mL). Mix the contents, add concentrated HCl (2.0 mL) and immediately stopper the flasks. In the third flask, prepare a blank sample by taking bromate-bromide solution (10.0 mL), distilled water (20.0 mL) and concentrated hydrochloric acid (2.0 mL) (do not add phenol soluton). Immediately stopper the flask and allow all the three flasks to stand for 25 minutes in water bath at room temperature. Shake the contents occasionally. A white colured precipitate of 2,4,6-tribromophenol begins to settle down in the test samples and the flasks acquire a yellow-brown colouration due to liberation of iodine. After 25 minutes, add potassium iodide solution (5.0 mL, 10%) to all the three flasks and immediately replace the stopper again. Allow the contents to stand for another 10 minutes. Rinse the stopper and neck of the flask with water (5.0 mL) and allow all the washings to run into the respective flasks. Titrate the test samples one by one with standardized sodium thiosulphate solution till straw yellow coloured solution is obtained. Add starch solution (5-6 drops), immediately observe blue-black colour of the solution. Continue addition of sodium thiosulphate till the solution becomes colourless. Note the volume of thiosulphate solution used with the test samples. Take the average of the two volumes (V_1 mL). Repeat the same procedure with the blank sample and note the volume of the sodium thiosulphate used (V mL).

Observations and Calculations:

Same as aniline (See *section 1.2.2*)

$$w = \frac{(V - V_1) \times N \times M}{1000 \times 6} \text{ g of phenol in 10.0 mL solution}$$

Strength of phenol in the given solution $= w \times 100$ g/L

$$\% \text{ Purity of phenol sample} = \frac{(V - V_1) \times N \times M}{1000 \times 6 \times W} \times 100$$

where, W = mass of phenol dissolved in 10.0 mL. (Divide the amount of phenol dissolved in one litre, by 100.)

Result: The strength of the given solution of phenol was found to be ------------ g/L.

Utility of This Exercise

See *section 1.2.2*

Lessons to Take Home

See *section 1.2.2*

1.4   ESTIMATION OF SUGARS

Sugars can be estimated by one of the following three methods:
- Chemical method depending upon the reducing properties
- Polarimetric method
- Fermentation method

Polarimetric method is the most accurate of the three methods listed above. However, the chemical method finds wide application in estimation of sugars in biological fluids. It rules out possible errors in polarimetric method due to the presence of other substances having optical rotation.

This section discusses the chemical method of sugar estimation.

1.4.1 Estimation of the Amount of Glucose Present in the Given Sample by Using Fehling's Solution

Basic Lab Operations: Titration

Estimated Time: 1 hour

Chemicals Required: Glucose, Fehling solutions A and B.

What You Should Know: Reducing sugars like glucose and mannose can be estimated volumetrically by using Fehling's solution (1:1 mixture of Fehling A and Fehling B). When equal volumes of Fehling A (aqueous copper sulphate solution with traces of sulphuric acid) and Fehling B (alkaline solution of Rochelle salt) are mixed, deep blue coloured copper tartrate is formed which keeps copper ions in solution.

A solution of reducing sugar on warming with copper tartrate gives red precipitate of cuprous oxide. The volume of Fehling's solution consumed for the fixed volume of sugar can be used for estimating the amount of sugar present in solution.

The reduction is not stoichiometric, therefore all the methods of estimation of reducing sugars are empirical and results are affected by slight variation in the procedure. The results are however trustworthy, if the standardization is done under identical conditions using pure sugar solutions.

Methylene blue can be used as an internal indicator towards the end of the reaction. The blue coloured dye is reduced to colourless compound in the presence of excess of reducing sugar. The indicator is so sensitive that the end point can be determined to within one drop of sugar solution in most cases.

Chemical Reaction Involved

$$CuSO_4 + 2NaOH \longrightarrow Cu(OH)_2 + Na_2SO_4$$

$$Cu(OH)_2 + 2 \begin{array}{c} COONa \\ | \\ H—C—OH \\ | \\ H—C—OH \\ | \\ COOK \end{array} \longrightarrow [\text{Copper complex}]^{2\ominus} + 2H_2O$$

Copper complex with sodium potassium tartrate (soluble)

$$\begin{array}{c} CHO \\ | \\ (CHOH)_4 \\ | \\ CH_2OH \end{array} + 2Cu^{2\oplus} + 5OH^{\ominus} \longrightarrow \begin{array}{c} COO^{\ominus} \\ | \\ (CHOH)_4 \\ | \\ CH_2OH \end{array} + Cu_2O\downarrow + 3H_2O$$

Glucose   **Gluconate**   **Brick red ppt.**

Experimental Procedure: Prepare a standard solution of glucose (0.044 M, 100 mL) by transferring ~0.8 g of anhydrous glucose (AR) to 100 mL standard flask. Add minimum amount of distilled water to dissolve the solid and make up the volume upto the mark. Prepare a mixture of Fehling's A and B (100 + 100 mL) in a dry beaker. Transfer this mixture to a burette. Run down, Fehling's solution (20.0 mL) from the burette into a dry conical flask and boil well on a wire gauze. Fill up another burette with standard glucose solution. Add glucose solution to the conical flask in the lots of 1.0 mL and shake. Ensure that the deep blue coloured solution of copper tartrate is boiling all the time. Continue the addition of glucose solution to the conical flask till the supernatant liquid becomes colorless. The titre value gives the approximate volume of glucose required for complete reduction of Fehling's solution (20.0 mL). Now repeat the same procedure and add so much of glucose solution that 1.0 mL of glucose is required for complete reduction. Boil the solution on the wire gauze, allow gentle ebullition for 2 minutes and complete the titration by adding glucose in lots of 0.1 mL, with shaking, until the supernatant liquid becomes colourless. The titre value now gives the exact volume of glucose required for complete reduction of 20.0 mL of the Fehling's solution. Repeat the procedure to obtain three concordant values (not differing by more than 0.1 mL). Repeat the same procedure with the given solution of glucose.

Observations and Calculations

Mass of glucose taken for standard solution = w g (for 100.0 mL)

Volume of Fehling's solution ($A + B$) used in each titration = 20.0 mL

Volume of standard glucose solution used = V_1 mL

Volume of given solution of glucose used = V_2 mL

20.0 mL of given Fehling's solution = V_1 mL of standard glucose solution

$$= V_2 \text{ mL of given glucose solution} - \frac{w \times V_1}{100} \text{ g of glucose}$$

$$\therefore\ 1000 \text{ mL of given glucose solution} = \frac{w \times V_1 \times 1000}{100 \times V_2} \text{ g of glucose}$$

$$\therefore\ \textit{Strength of the given glucose solution} = \frac{w \times V_1 \times 10}{V_2} \text{ g/L}$$

Determination of percentage purity

If W = amount of glucose dissolved in 1000 mL of given glucose solution

Then,

$$\textit{\% Purity of the given sample of glucose} = \frac{w \times V_1 \times 10 \times 100}{V_2 \times W}$$

Note

- Always use a bulb sucker or a burette to measure out the volume of Fehling A, as copper sulphate is highly toxic and poisonous and Fehling solution B is strongly alkaline.

- For preparation of Fehling solutions A and B: See Appendix II
- Fehling solution deteriorates on standing, due to precipitation of cupric hydroxide. Therefore, it should be prepared fresh, whenever required, by mixing Fehling's solution A and B.
- While adding glucose solution into the boiling Fehling's solution, it is advisable to allow the solution to cool from time to time, permitting the cuprous oxide precipitate to settle. The solution in the flask can then be tilted on one side and the supernatant liquid can be seen on the white background of water basin. This enables proper viewing of the colour of the supernatant liquid. Do not forget to bring the solution to boil before resuming the addition of glucose solution.
- Clearly, this estimation can be used for reducing sugars only.
- The reduction in this experiment is however not stoichiometric, nevertheless the results are trustworthy if the experiment details are adhered to and if standardization is affected under similar conditions.
- It is advisable, to determine the approximate end point in the first addition as described above, by adding glucose in lots of 1.0 mL. For determination of exact end point, add glucose in lots of 1.0 mL about 0.5 mL-1.0 mL before the expected end point and then add 3-5 drops of 1% methylene blue to the hot boiling solution. Allow the solution to boil for another 1-2 minutes. Continue addition of the remaining glucose solution dropwise until a colourless solution is obtained. This method allows accurate end point determination within one drop of glucose solution. Repeat the titration to get three concordant values.
- Aqueous solution of methylene blue may be prepared by dissolving 1.0 g of methylene blue in 100 mL of distilled water.

Result: The strength of the given solution of glucose was found to be -------- g/L.

Utility of This Exercise

- Determination of the amount of reducing sugars in biological fluids.

Lessons to Take Home

After doing this exercise, the students will be able to do the following:

- Estimate the percentage purity of the commercial samples of reducing sugars.
- Estimate the amount of reducing sugar in the biological fluids.
- Understand why sucrose cannot be estimated by this exercise whereas maltose and lactose can be.
- Understand why it is necessary to keep the Fehling's solution boiling while adding glucose from the burette.

1.5 DETERMINATION OF SAPONIFICATION VALUE OF THE GIVEN SAMPLE OF FAT OR OIL

Basic Lab Operations: Refluxing, titration.

Estimated Time: 90 minutes

Chemicals Required: Oil/fat, alcoholic potassium hydroxide, oxalic acid, phenolphthalein.

What You Should Know: Naturally occurring fats/oils are triglycerides or triesters of glycerol with fatty acids (C-12 to C-20 even chain carboxylic acids). Those triglycerides which contain higher proportion of unsaturated fatty acid component are liquids at room temperature and are called **oils**. All edible oils like olive oil, sunflower oil, groundnut oil and soyabean oil contain unsaturated fatty acids and are thus liquids at room temperature.

Saponification value of a fat or oil is defined as number of milligrams of potassium hydroxide required to completely hydrolyse one gram of fat or oil. The process leads to formation of potassium salts of fatty acids (called **soaps**), which have detergent properties, hence, is called *saponification*.

Chemical Reaction involved

$$
\begin{array}{l}
CH_2OCOR' \\
| \\
CHOCOR'' \quad + 3KOH \longrightarrow \\
| \\
CH_2OCOR'''
\end{array}
\quad
\begin{array}{l}
CH_2OH \\
| \\
CHOH \quad + R'COOK \quad + R'''COOK \quad + R''COOK \\
| \\
CH_2OH
\end{array}
$$

Triglyceride (Fat/Oil)	Glycerol	Soaps

Experimental Procedure: Prepare alcoholic potassium hydroxide solution (200 mL, 0.5 N) in a volumetric flask. Replace the stopper and keep the solution overnight. Filter any insoluble precipitate before use.

Standardize the alcoholic potassium hydroxide solution by titrating against standard oxalic acid solution (0.5 N), using phenolphthalein as indicator. Take the given sample of fat/oil (1.0 g) in a round bottomed flask (100 mL). Add standardized alcoholic potassium hydroxide solution (50.0 mL). Reflux on the steam bath for 1.5 to 2 hours, using a reflux water condenser, or till the sample shows no droplets of oil. Cool and titrate the solution with standard oxalic acid using phenolphthalein as the indicator. Record the volume of oxalic acid used as V_2. Carry out a blank titration of alcoholic KOH (50.0 mL) with standard oxalic acid and record the volume as V_1.

Note

- Saponification is normally carried out in alcoholic alkali hydroxide as oils are insoluble in water.

- For preparation of alcoholic potassium hydroxide solution (0.5 N): See Appendix II.
- If ground glass joints are used during refluxing, the flask should be removed from the condenser immediately after refluxing is over. This will prevent the sticking of joints due to the presence of alkali.
- Transesterification of the oil with methanolic or ethanolic alkali hydroxide to form volatile methyl or ethyl esters cannot be ruled out during saponification. The two esters exist in equilibrium before being hydrolyzed.
- Ensure that the contents are free from any suspended oily droplets before removing the R.B. flask from steam bath.

Observations and Calculations

Weight of oil/fat = w g

Volume of oxalic acid used for hydrolysis of sample

$$= V_1 - V_2 \text{ mL}$$

Volume of KOH solution used for hydrolysis of sample = V mL

$$= \frac{(V_1 - V_2) \times N_{\text{oxalic acid}}}{N_{\text{KOH}}}$$

$$1000.0 \text{ mL of } 1N \text{ KOH} = 56.0 \text{ g of KOH}$$

$\therefore \qquad 1000.0$ mL of N normal KOH $= 56 \times N_{\text{KOH}}$ g of KOH

$$V \text{ mL of } N \text{ normal KOH} = \frac{56 \times N_{\text{KOH}} \times V}{1000} \text{ g of KOH}$$

= Amount of KOH in grams used for w g of oil

$\therefore$ Saponification value of the oil (S) $= \dfrac{56 \times V \times N_{\text{KOH}}}{w}$ mg/g

M g of fat will require 168000 mg of KOH (See chemical reaction)

$\therefore \qquad$ 1.0 g of fat will require $= \dfrac{168000}{M}$ mg of KOH $= S$

$$\text{Molecular weight of oil } (M) = \frac{3 \times 56 \times 1000}{S}$$

Result: The saponification value of the given sample of fat/oil was found to be ----------.

Utility of This Exercise

- This exercise is useful for the analysis of animal or vegetable fat or oil sample.
- Saponification values are used to determine the amount of alkali required to completely saponify the given sample of fat or oil during

commercial preparation of soaps. This ensures that the end product does not contain excess of alkali, as the same can be harmful for the skin.

Lessons to Take Home

After doing this exercise, the students will be able to do the following:

- Predict the exact amount of KOH required for saponification.
- Determine average molecular weight of an oil/fat. Higher molecular weight means smaller saponification value.
- Determine the nature of fatty acid in the fat or oil. Long chain fatty acid esters have lower saponification value because fewer COOH groups are present/unit M of the sample.
- Predicting adulteration of petroleum oils. Petroleum oils do not saponify whereas animal fat and vegetable oils do. So, the saponification value indicates adulteration in blended lubricating oils.

1.6 DETERMINATION OF IODINE VALUE OF THE GIVEN SAMPLE OF FAT OR OIL

Basic Lab Operations: Iodometric titration.

Chemicals Required: Oil/fat, Wij's solution, potassium iodide, potassium dichromate, sodium thiosulphate, sulphuric acid, starch solution, glacial acetic acid and iodine trichloride.

What You Should Know: Iodine value of a fat or oil is defined as the number of grams of iodine required to completely saturate 100 grams of fat or oil. This process leads to addition of iodine across the $-C=C-$. Iodine value of zero represents a completely saturated fat, like coconut oil.

The given sample of oil or fat is treated with the known excess of Wij's solution (ICl, V mL), which adds across the $-C=C-$ in the glyceride fragment. The unreacted ICl is converted to equivalent amounts of iodine, with potassium iodide, which is then estimated volumetrically by titration with standardized solution of sodium thiosulphate (V_2 mL). A simultaneous blank titration is performed to determine the total amount of iodine evolved from the V mL of ICl (V_1 mL).

Chemical Reaction Involved

$$-C=C- \; + \; ICl \longrightarrow -\overset{|}{\underset{|}{C}}-\overset{|}{\underset{|}{C}}- $$
$$ I \quad Cl$$

$$ICl \; + \; KI \longrightarrow I_2 \; + \; KCl$$

$$I_2 \; + \; 2Na_2S_2O_3 \longrightarrow 2NaI \; + \; Na_2S_4O_6$$

Experimental Procedure: Prepare sodium thiosulphate solution (~ 0.1N, 100 mL), standard potassium dichromate solution (0.1 N, 50 mL), 15% potassium iodide solution (100 mL) and keep them aside.

Next, prepare a solution of iodine trichloride (7.9 g) in glacial acetic acid (100.0 mL), in a conical flask, by warming on a water bath. In another flask take iodine (8.7 g) and dissolve in another 100.0 mL of warm glacial acetic acid. Mix the two solutions in a 1000 mL volumetric flask and make up the volume with glacial acetic acid. Store the resulting Wij's solution in a stoppered amber coloured bottle. (This solution is stable for one month!)

Take the sodium thiosulphate solution in the burette, pipette out standard potassium dichromate solution (10.0 mL) in a clean conical flask. Add sulphuric acid (1N, 10.0 mL), potassium iodide (10.0 mL, 15%). Cover the conical flask and keep it undisturbed for 10 minutes. The solution becomes dark yellow due to liberated iodine. Titrate the liberated iodine against sodium thiosulphate solution till solution becomes straw yellow. Add 5-6 drops of starch and continue addition of sodium thiosulphate solution till the solution becomes light green. Repeat to get three concordant values. Determine normality of the sodium thiosulphate solution.

Next, take the given sample of oil (0.2-0.3 g of peanut oil or corn oil or cotton seed oil) in an iodine flask (250 mL). To this add CCl_4 (25-30 mL), warm, if required, on water bath. Add Wij's solution (25.0 mL), stopper the flask and shake for a few minutes. Allow the flask to stand in dark for 30 minutes, shake the contents occasionally. Add KI solution (20.0 mL; 15%) and dilute the mixture with water (100 mL). Titrate the mixture immediately against 0.1 N sodium thiosulphate solution, till the colour turns pale yellow. Add 5-6 drops of starch, the colour will immediately turn blue. Continue the addition of sodium thiosulphate till the colour turns colourless. Perform a simultaneous blank titration with Wij's solution (25.0 mL). Except oil, add all reagents as in test sample while carrying out the blank titration.

Observations and Calculations

Weight of oil/fat $= w$ g

If, $V_1 =$ volume of $Na_2S_2O_3$ used with the blank

$M =$ equivalent mass of iodine $= 127$ g

$V_2 =$ volume of $Na_2S_2O_3$ used with the sample

$N =$ normality of $Na_2S_2O_3$ solution

Then, volume of sodium thiosulphate used by the sample $= V_1 - V_2$

1000 mL of 1N $Na_2S_2O_3$ = 1000 mL of 1N iodine = 127.0 g of iodine = M g of iodine

$\therefore$ $V_1 - V_2$ mL of N normal $Na_2S_2O_3$ = $V_1 - V_2$ mL of N normal iodine solution

$$= (V_1 - V_2) \times N \times \frac{M}{1000} \text{ g of iodine}$$

$$\therefore \qquad w \text{ g of fat or oil} = (V_1 - V_2) \times N \times \frac{M}{1000} \text{ g of iodine}$$

$$100 \text{ g of fat or oil} = (V_1 - V_2) \times N \times \frac{M \times 100}{1000 \times w} \text{ g of iodine}$$

Therefore,

$$\textit{Iodine value of given fat/oil} = (V_1 - V_2) \times N \times \frac{M}{10 \times w} \text{ g of iodine}$$

Result: The iodine value of the given sample of fat/oil was found to be
-------------.

Utility of This Exercise

- Iodine values are used to check the quality control parameters for traded oil products

Lessons to Take Home

After doing this exercise, the students will be able to do the following:

- Understand why Wij's solution should be used in excess.
- Predict the degree of unsaturation in a given sample of oil. Higher iodine value means greater unsaturation.
- Categorize the oils into drying, semi-drying and non-drying oils.

1.7 DETERMINATION OF THE STRENGTH OF THE GIVEN SOLUTION OF GLYCINE BY SORENSEN FORMOL TITRATION

Basic Lab Operations: Titration

Estimated Time: 1 hour

Chemicals Required: Glycine, formalin, sodium hydroxide, phenolphthalein and hydrochloric acid.

What You Should Know: Amino acids are characterized by the presence of one or more acidic (–COOH) and basic groups (–NH$_2$, guanidine group or imidazole ring). They exist in one of the three forms, *i.e.*, conjugate acid/ conjugate base/Zwitter ion in solution, depending upon the pH.

Estimation of the amount of amino acid in a solution can be carried out by either estimating the free amino group (**Van Slyke method**) or by estimating the free carboxyl group (**Sorensen Formol Titration**). Estimation of the free carboxyl group cannot be carried out directly by titration with sodium hydroxide solution as the acidic group for the amino acid in solution is ammonium ion and not –COOH group. Further, buffering action of the zwitterion prevents accurate estimation. Therefore, the amino acid is first treated with carefully neutralized formaldehyde to block the amino group as methylene or dimethylol derivative, before titration with standardized sodium hydroxide solution. Since the formalin solution contains formic acid and amino acid solution is also seldom neutral, it is important that both amino acid and formalin solution are brought to approximately same pH before mixing. For this purpose, each solution is made just alkaline to phenolphthalein by adding dilute sodium hydroxide solution. This preliminary neutralization should not however be confused with titration of neutralized glycine with sodium hydroxide solution. Most amino acids in water exist as zwitter ion. Therefore, they cannot react with formaldehyde solution directly. This problem can be solved if the amino acid solution is first treated with just enough sodium hydroxide to liberate the free amino group and then reaction with formalin is carried out. The free carboxyl group can then be estimated by titration with standardized NaOH solution. The overall reaction may be represented in the following manner, although the precise reactions are more complicated.

Chemical Reaction Involved

$$\overset{\oplus}{H_3N}-\overset{\overset{\displaystyle H}{|}}{\underset{\underset{\displaystyle R}{|}}{C}}-COOH \underset{}{\overset{pK_1}{\rightleftharpoons}} \overset{\oplus}{H_3N}-\overset{\overset{\displaystyle H}{|}}{\underset{\underset{\displaystyle R}{|}}{C}}-COO^{\ominus} \underset{}{\overset{pK_2}{\rightleftharpoons}} NH_2-\overset{\overset{\displaystyle H}{|}}{\underset{\underset{\displaystyle R}{|}}{C}}-COO^{\ominus}$$

Conjugate Acid   **Zwitterion**   **Conjugate Base**

$$\overset{\oplus}{H_3N}CH_2COO^{\ominus} + NaOH \rightleftharpoons H_2NCH_2COONa + H_2O$$

$$\downarrow CH_2O$$

$$(CH_2OH)_2NCH_2COONa \ + \ H_2C=NCH_2COONa$$

Dimethylol derivative   **Methylene derivative**

Experimental Procedure: Prepare N/10 standard glycine solution by weighing out accurately glycine (0.75 g) and dissolving it in distilled water (100 mL). Take formalin solution (30-40% commercial sample, 50.0 mL) in a conical flask (with the help of a measuring cylinder) and add 2-3 drops of phenolphthalein. Add sodium hydroxide solution (0.1 N) from the burette, dropwise, till the solution turns pink. This is done to neutralize the otherwise acidic solution of formalin. Keep this solution aside.

Pipette out standard glycine solution (20.0 mL) in a conical flask and add 2-3 drops of phenolphthalein. Add sodium hydroxide solution from the burette till the colour just becomes light pink. (This is neutralized glycine!) To this add neutralized formalin solution (10.0 mL). Note that the pink colour disappears immediately and the solution becomes markedly acidic. Note the reading on the burette and then titrate this solution with 0.1N sodium hydroxide solution. Repeat this procedure to obtain three concordant readings (V_1 mL).

Repeat the procedure with the given solution of glycine.

Observations and Calculations

Mass of glycine transferred in standard solution = w g

$$\text{Normality of standard glycine solution } (N_1) = \frac{w \times 1000}{75 \times 100}$$

Volume of NaOH used for standard glycine solution = V_1 mL

$$\text{Normality of NaOH } (N_2) = \frac{N_1 \times 20}{V_1} N$$

Volume of NaOH used for the given glycine solution = V_2 mL

$$\text{Normality of the given glycine solution } (N_3) = \frac{N_2 \times V_2}{20} N$$

Strength of the given glycine solution = Normality × equivalent mass
= $N_3 \times 75$ g/L

$$\textit{Percentage Purity of Glycine} = \frac{\text{Strength g/L}}{w \times 10} \times 100\%$$

Result: The strength of the given solution of glycine was found to be ---------- g/L.

Utility of This Exercise

- This method is used for following the course of enzymatic hydrolysis of a protein/peptide. For example, if a peptide hydrolysis is represented as follows:

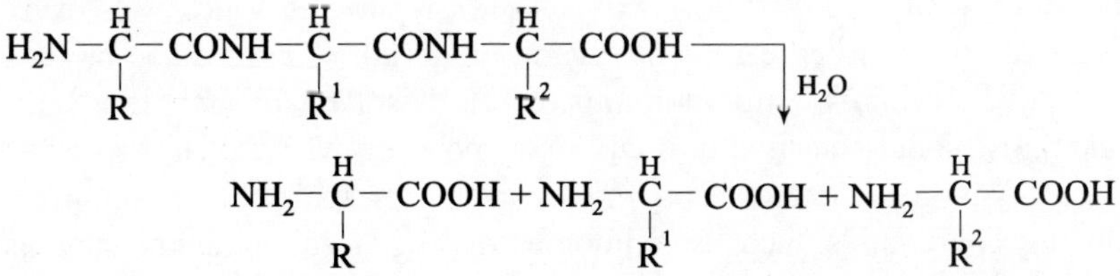

$$H_2N - \underset{R}{\overset{H}{C}} - CONH - \underset{R^1}{\overset{H}{C}} - CONH - \underset{R^2}{\overset{H}{C}} - COOH \xrightarrow{\;H_2O\;}$$

$$NH_2 - \underset{R}{\overset{H}{C}} - COOH + NH_2 - \underset{R^1}{\overset{H}{C}} - COOH + NH_2 - \underset{R^2}{\overset{H}{C}} - COOH$$

- It is clear that if the above estimation is carried out on original peptide and its hydrolysed product, the latter will require three times as much NaOH as the former.

Lessons to Take Home

After doing this exercise, the students will be able to do the following:

- Understand why formalin and glycine solutions need to be neutralised prior to titration.

- Understand why carboxylic group of an amino acid (or Sorensen's Formol titration) cannot be estimated directly by titrating the aqueous solution of amino acid with aqueous alkali.
- Estimate the amount of amino acid in a given solution.
- Determine percentage purity of a commercial sample of amino acid.

1.8　DETERMINATION OF THE AMOUNT OF DISSOLVED OXYGEN (DO) IN THE GIVEN WATER SAMPLE BY MODIFIED WINKLER'S METHOD

Basic Lab Operations: Iodometric titration.

Estimated Time: 2 hours

Chemicals Required: Manganese sulphate monohydrate, sulphuric acid (4N), sodium hydroxide, sodium azide, starch, potassium iodide, sodium thiosulphate, potassium dichromate.

What You Should Know: The amount of free oxygen dissolved in water is called **dissolved oxygen** (DO). It is expressed in mg/L or ppm. The main sources of oxygen in water are: (a) photosynthetic activity occurring on the surface of water, (b) absorption/diffusion. The maximum solubility of oxygen in water at 1 atmosphere pressure ranges from about 15 mg/L at 0°C to 8 mg/L at 30°C. DO levels in natural and wastewater vary with the temperature of air and water, degree of hardness of water, demand of oxygen in the water body and altitude. Cold water holds more oxygen than warm water. Thermal discharges, such as water used as coolant in power plants, raise the temperature of water and lower its oxygen content. Solubility of oxygen also decreases with altitude.

DO is an important water quality parameter since it is essential for a healthy aquatic ecosystem. Fish and aquatic animals need the oxygen dissolved in the water to survive. Low levels of oxygen (hypnoxia) or no levels of oxygen (anoxia) can occur when excess of organic material, such as algae blooms, are decomposed by aerobic microorganisms. If dissolved oxygen levels decline substantially, some sensitive animals may move away, weaken, or die. In the absence of oxygen, anaerobic organisms take over the decomposition process producing compounds such as ammonia, hydrogen sulphide, mercaptans etc. These substances cause environmental nuisance. On the other hand, too much of oxygen in water is not good for the industrial use, as excess of oxygen corrodes the machines due to slow oxidation.

Dissolved oxygen in a water sample may be determined either by dissolved oxygen meter or by a titrimetric method called **Winkler's method**. Winkler's method is an iodometric method. In this method, manganese (II) sulphate and alkaline iodide-azide reagent are added to the water sample. Manganese (II) sulphate in alkaline medium gets precipitated as white coloured, $Mn(OH)_2$. The DO present in water oxidises Mn(II) hydroxide, giving a brown precipitate of manganese dioxide, MnO_2. The amount of MnO_2 produced in the sample is proportional to the amount of dissolved oxygen. MnO_2 oxidises the iodide ions present in the solution to iodine and is itself reduced to soluble manganese sulphate, in the acidic medium. The liberated iodine is estimated volumetrically by titrating it against standardized sodium thiosulphate solution. The end point in the titration is reached when the blue coloured starch-iodine complex disappears and a colourless solution is obtained. Therefore, technically the volume of sodium thiosulphate used is equal to the amount of dissolved oxygen in water.

$$MnSO_4 + 2\,NaOH \longrightarrow Mn(OH)_2 + Na_2SO_4$$

$$2\,Mn^{2\oplus} + 4\,OH^{\ominus} + O_2 \longrightarrow 2\,MnO_2 + 2\,H_2O$$

$$MnO_2 + 4\,H^{\oplus} + 2\,I^{\ominus} \longrightarrow Mn^{2+} + I_2 + 2\,H_2O$$

Sodium thiosulphate solution may be standardised by iodometric titration against standard potassium dichromate solution as per the following reaction:

$$6I^{\ominus} + Cr_2O_7^{2\ominus} + 14\,H^{\oplus} \longrightarrow 3\,I_2 + 2\,Cr^{3\oplus} + 7\,H_2O$$

$$I_2 + 2\,S_2O_3^{2\ominus} \longrightarrow 2\,I^{\ominus} + S_4O_6^{2\ominus}$$

Effluents from wastewater treatment plants employing biological processes, river water and incubated BOD samples contain nitrite ions. These nitrite ions oxidize iodide ions to molecular iodine under acidic conditions. Further, the reduced form of nitrite (N_2O_2) is oxidized by oxygen to NO_2^- ion again, establishing a cycle that can result in erroneous results, far in excess of amounts that would be expected. Winkler has proposed alkali-iodide azide modification where, sodium azide (NaN_3) is added along with the alkali-KI reagent to the water sample. This destroys the nitrite ions as per the reaction given below and enables accurate determination of DO:

$$2\,NO_2^{\ominus} + 4\,H^{\oplus} + 2\,I^{\ominus} \longrightarrow N_2O_2 + I_2 + 2\,H_2O$$

$$NaN_3 + H^{\oplus} \longrightarrow HN_3 + Na^{\oplus}$$

$$HN_3 + NO_2^{\ominus} + H^{\oplus} \longrightarrow N_2O + N_2 + H_2O$$

Experimental Procedure: Prepare a solution of potassium dichromate (0.01 N) and sodium thiosulphate (~0.01N). Take sodium thiosulphate in the burette and note the initial reading. Pipette out standard potassium dichromate (10.0 mL) into a clean titration flask, add dilute sulphuric acid (4N, 5.0 mL) and potassium iodide (5%, 5.0 mL). Cover the flask with ordinary filter paper and keep the solution aside for 5-10 minutes. Now, add sodium thiosulphate solution from the burette till the solution turns straw yellow. Add 5-6 drops of starch and notice the blue-violet color of the starch-iodine complex in the solution. Continue adding sodium thiosulphate solution until the colour changes from blue to light green. Repeat this procedure to get three concordant values. Determine the normality of sodium thiosulphate solution.

Collect the water sample to be tested in a 300 mL BOD bottle taking care to avoid adding air to the water being collected. This is done by ensuring that the bottle remains completely submerged during filling and the stopper is inserted holding the filled bottle under water. Remove water (3.0 mL) from the bottle with the help of a pipette. Use auto pipette to add manganese(II) sulphate solution (1.0 mL) and alkali-iodide-azide solution (1.0 mL) to the water sample. (Do not pipette with mouth!) Stopper and invert the bottle several times to mix thoroughly. Allow the solid formed to settle to about half the volume of the bottle. Invert the bottle several times to mix the solid back into the solution. Allow the solid formed to settle again. After settling, add concentrated sulphuric acid (1.0 mL), stopper and gently invert several times. Continue inverting the flask until the precipitate redissolves in the solution.

Take 50.0 mL of the sample in a 250 mL conical flask (*use burrete!*) and titrate against standardized sodium thiosulphate solution until the colour turns pale yellow. Add starch (1.0 mL) and continue titration until the blue/violet colour is discharged. Repeat the whole process with another sample of water collected from the same place. Take average of the volumes of sodium thiosulphate used with the test samples (V mL).

Observations and Calculations

1000 mL of $1N$ $Na_2S_2O_3$ = 8.0 g of oxygen

$$\therefore V \text{ mL of } N \text{ normal } Na_2S_2O_3 = 8 \times V \times \frac{N}{1000} \text{ g of oxygen}$$

$$50.0 \text{ mL of the sample contains} = 8 \times V \times \frac{N}{1000} \times 1000 \text{ mg of oxygen}$$

$$\therefore 1000 \text{ mL of the sample contains} = 8 \times \left\{ V \times N \times \frac{1000}{50} \right\} \text{ mg of oxygen}$$

V = Volume of sodium thiosulphate used, equivalent mass of oxygen = 8

N = Normality of sodium thiosulphate

$$\therefore DO = \frac{\text{Titre value} \times \text{Normality of titrant} \times 8 \times 1000}{\text{Volume of sample titrated}} \text{ mg/L}$$

Note

- Success of this method depends upon the manner in which the sample of water is manipulated.
- At all stages care must be taken to ensure that oxygen is neither introduced nor lost from the sample.
- For accurate DO determination, water samples should be free from any solutes that oxidise iodide ion or reduce iodine.

- If no precipitate is formed on adding manganese sulphate and alkali-iodide-azide, it indicates that the sample has no dissolved oxygen. In such cases, it is appropriate to obtain another sample and ensure that no error is introduced in sample preparation.
- **Manganese sulphate solution**: For preparation, see Appendix II
- **Alkali-iodide-azide**: For preparation, see Appendix II

Result: The DO of the given water sample was found to be......................... mg/L.

Utility of This Exercise

- DO is an important water quality parameter since it is essential for a healthy aquatic ecosystem. It can be used to determine the amount of dissolved oxygen in uncontaminated oceans.
- It can be used by marine chemists for determining the redox potential of a water column.

Lessons to Take Home

After doing this exercise, the students will be able to do the following:

- Determine the health and cleanliness of a lake or stream.
- Know about the amount and type of biomass a fresh water sample can support.
- Estimate the amount of decomposition occurring in a lake or stream.

1.9 DETERMINATION OF BIOCHEMICAL OXYGEN DEMAND IN THE GIVEN WATER SAMPLE

Basic Lab Operations: Iodometric titration.

Estimated Time: 5 days

Chemicals Required: Manganese sulphate, sulphuric acid, sodium hydroxide, sodium azide, starch, potassium iodide, sodium thiosulphate and potassium dichromate.

*What You Should Know***:** Determining how organic matter affects the concentration of dissolved oxygen is integral to water quality management across the world. Biochemical oxygen demand (**BOD**) is a test which determines the amount of organic material in wastewater by measuring the oxygen consumed by microorganisms in decomposing organic constituents of the waste. The higher the BOD, the more oxygen will be demanded from the waste to break down the organics. The BOD test is performed by incubating a sealed water sample for the standard 5-day period, then determining the change in dissolved oxygen content.

Experimental Procedure: Collect 300 mL each of the water sample to be tested in two BOD bottles. Carry out DO analysis from one bottle using the procedure described in *section 1.8* for day 0 (*i.e.*, Sample 0 Day) and keep the other sample in BOD incubator at 20°C for 5 days. Carry out DO analysis with this water sample after five days of incubation. Then, calculate the BOD value of the given water sample by the following equation:

$$\text{BOD of water sample} = \frac{\text{DO of water sample at 0 day}}{\text{DO of water sample at 5 day}}$$

Result: The BOD of the given water sample was found to be .... mg/L

Utility of This Exercise

- BOD is used often in wastewater treatment plants (WTP) as an index of degree of organic pollution in water.
- Efficiency of WTP can be predicted by this exercise.

Lessons to Take Home

After doing this exercise the students will be able to do the following:

- Understand the terms DO and BOD and significance of their determination.
- Understand why DO in a given water sample deceases with time.

1.10 DETERMINATION OF CHEMICAL OXYGEN DEMAND (COD) IN A GIVEN WATER SAMPLE

Basic Lab Operations: Titration

Estimated Time: 90 minutes

Chemicals Required: Potassium dichromate, Mohr's salt, silver sulphate, mercuric sulphate, Ferroin, sulphuric acid.

What You Should Know: Chemical oxygen demand (**COD**) is a measure of the oxygen equivalent of the portion of organic waste, present in wastewater, that can be oxidized by a strong oxidizing agent like potassium dichromate. It is an easily determined parameter and its value is reported in mg/L. COD is always higher than the biochemical oxygen demand (BOD), since it includes both organic matter which is degradable and non-degradable biologically.

There are two methods for COD determination: (a) Open reflux method which is suitable for a wide range of wastewater and is preferred for large sample size. (b) Closed reflux method, which is suitable for smaller quantities and requires homogenization of the sample. In the closed reflux method, water sample is refluxed with excess of oxidizing agent like acidified potassium dichromate. Some of the dichromate is used for oxidation of the organic matter and the unreacted excess is estimated by titration with standard Mohr's salt solution. The amount of dichromate used for the oxidation of the organic matter in water is equivalent to the amount of oxygen that would be required for the same purpose. The reaction between potassium dichromate and Mohr's salt solution is given below:

$$K_2Cr_2O_7 + 6FeSO_4(NH_4)_2\ SO_4.6H_2O + 7H_2SO_4 \longrightarrow$$
$$Cr_2(SO_4)_3 + K_2SO_4 + 3Fe_2(SO_4)_3 + 6(NH_4)_2SO_4 + 7H_2O$$

Experimental Procedure: Place 20.0 mL each of the water sample to be tested and distilled water in two round bottomed flasks (A and B) respectively. Add a pinch of mercuric sulphate, silver sulphate, potassium dichromate solution (10.0 mL, 0.25 N) and dilute sulphuric acid (2N, 30.0 mL) to both the flasks and reflux the contents for 2 hours. Allow the reaction mixtures to attain room temperature and transfer the contents into two different conical flasks. Wash both the R.B. flasks with distilled water and transfer the washings to the respective conical flasks. Add 2-3 drops of ferroin to each and titrate against standard ferrous ammonium sulphate solution (0.25N), till the color changes from blue-green to reddish-brown. Note the volume of Mohr's salt solution used for both samples, **A** and **B**.

Volume of standard ammonium ferrous sulphate solution used for blank (**B**) = V_1 mL

Volume of standard ammonium ferrous sulphate solution used for test sample (**A**) = V_2 mL

Normality of standard ammonium ferrous sulphate solution = N

Volume of water sample taken = V_3 mL

Equivalent mass of oxygen = 8.0

1000 mL of $1N$ Mohr's salt = 8.0 g of oxygen

$$\therefore\quad V_1 - V_2 \text{ ml of } N \text{ normal Mohr salt} = \frac{8 \times (V_1 - V_2) \times N}{1000} \text{ g of oxygen}$$

$$20.0 \text{ mL of water sample} = \frac{8 \times (V_1 - V_2) \times N \times 1000}{1000} \text{ mg of oxygen}$$

$$\therefore\quad 1000 \text{ mL of water sample} = \frac{8 \times (V_1 - V_2) \times N \times 1000}{20} \text{ mg of oxygen}$$

$$\therefore\qquad COD = \frac{(V_1 - V_2) \times N \times 8 \times 1000}{\text{Volume of water sample taken}} \text{ mg/L}$$

Note

- Silver sulphate is used as catalyst in the reaction.
- Mercuric sulphate is added to scavenge chloride ions from water sample.
- This method cannot differentiate between the biologically inert and oxidizable material.

- Potassium dichromate solution (N/4): Dissolve potassium dichromate (1.22 g) in distilled water (100.0 mL).
- Mohr's salt solution (N/4): In a 100 mL standard flask, dissolve Mohr's salt (9.8 g) in minimum amount of distilled water, add dilute sulphuric acid (8.0 mL) and make up the volume to 100 mL.
- Ferroin reagent: See Appendix II

Result: The COD of the given water sample was found to be......................... mg/L.

Utility of This Exercise

- COD can be used to determine degree of pollution in water bodies.
- It can estimate the efficiency of wastewater treatment plants (WWTP).

Lessons to Take Home

After doing this exercise, the students will be able to do the following:

- Understand the meaning and utility of COD.
- Understand how BOD is related to COD.
- Understand why COD values are higher than BOD values.
- Get a estimate of the COD in a water sample.
- Determine the pollution loads in a water body

1.11 ISOLATION AND ESTIMATION OF THE AMOUNT OF ASPIRIN IN A COMMERCIAL TABLET

Basic Lab Operations: Extraction and acid-base titration.

Estimated Time: 45 minutes

Chemicals Required: Aspirin tablets, ethanol, sodium hydroxide, phenolphthalein, oxalic acid.

What You Should Know: Aspirin is nothing but acetyl salicylic acid. Aspirin is an antipyretic, analgesic and anti-inflammatory drug. As an anti-inflammatory agent, aspirin is used extensively in the treatment of arthritis. Nowadays, it is universally recommended as a medicine which may prevent heart attack by checking blood clotting in the arteries.

The commercial tablets contain acetyl salicylic acid together with common binding substances including starch, methyl cellulose and microcrystalline cellulose. The presence of starch in aspirin tablet can be confirmed by boiling one fourth of the aspirin tablet with water (2.0 mL). The solution is cooled and a few drops of iodine solution are added. Formation of deep blue coloured starch-iodine complex confirms the presence of starch in the commercial sample.

Purity of any compound is important for its action. It is all the more important if the compound happens to be a drug. The purity of aspirin can be checked qualitatively as simple colour test in the lab. Development of brown colour with $FeCl_3$ indicates the presence of salicylic acid in the preparation. This salicylic acid originates from the hydrolysis of acetylsalicylic acid. This hydrolysis brings in another impurity, acetic acid. The smell of vinegar, experienced on opening an old bottle of aspirin tablets is due to acetic acid formed during hydrolysis.

The presence of salicylic acid, whether as an impurity in the preparation or consequence of hydrolysis, is not desirable. Incidentally, salicylic acid is also an analgesic but is not as safe as aspirin, because the free –OH group causes severe mucousal irritation and gastric problems. Excessive use may even cause gastric ulcers. Furthermore, the effectiveness of a drug depends on its proper dosage besides the purity, hence the amount of aspirin per tablet is normally marked on the packing. The amount of aspirin in any preparation can be determined by a number of methods including simple titrimetry, conductometry, potentiometry and colorimetry methods. This section describes the detailed principle and procedure for the titrimetric method.

To determine aspirin titrimetrically, it is first dissolved in 30% alcohol and then titrated with standardized solution of alkali. One equivalent of aspirin will neutralise one equivalent of NaOH due to the presence of free COOH group.

Experimental Procedure: Prepare a standard solution of oxalic acid (100 mL, N/50) and sodium hydroxide (N/100). Take the sodium hydroxide solution in the burette, pipette out oxalic acid solution (10.0 mL) in the conical flask and titrate it against sodium hydroxide solution taking phenolphthalein as the indicator. The end point is indicated by appearance of faint pink colour. Repeat the titration to get three concordant readings. Determine the normality of the sodium hydroxide solution.

Take 4 tablets of aspirin and finely powder them in a mortar and pestle. Dissolve the powder in aqueous ethanol (50%, 30.0 mL), warm if required. Titrate the resulting solution against standardized sodium hydroxide solution taking phenolphthalein as the indicator. Repeat the titration to get three concordant values.

Observations and Calculation

Mass of oxalic acid transferred = w g

Normality of standard oxalic acid solution $(N_1) = \dfrac{w \times 1000}{63 \times 100}$ N

Mass of Aspirin transferred = w_1 g

Volume of aspirin solution taken = 30.0 mL

Strength of aspirin solution prepared = $\dfrac{w_1}{30} \times 1000$ g/L

Volume of NaOH used for standard oxalic acid solution = V_1 mL

Normality of NaOH solution (N_2) = $\dfrac{N_1 \times 10}{V_1}$

Volume of NaOH used for given aspirin tablet solution = V_2 mL

Normality of aspirin solution (N_3) = $\dfrac{N_2 \times V_2}{30}$ N

Strength of aspirin solution = Normality × equivalent mass = $N_3 \times 180.158$ g/L

$$\% \ of \ Aspirin \ in \ the \ Tablet = \dfrac{N_3 \times 180.158}{w_1 \times 1000} \times 30 \times 100$$

$$= \dfrac{3 \times N_3 \times 180.158}{w_1}$$

Note

- The estimation assumes that aspirin sample is free from salicylic acid.

Result: The percentage of aspirin in the given tablet was found to be ....

Utility of This Exercise

This estimation is useul for the following:

- Determination of percentage purity of the commercial samples of aspirin, as a means of quality check.
- Ruling out the presence of salicylic acid in the commercial sample.

Lessons to Take Home

After doing this exercise, the students will be able to do the following:

- Understand why it is important to do purity checks for a drug before it is marketed.
- Carry out similar estimations on other drugs having acidic or basic groups.

Post Lab Questions

1. Name some common methods used for the estimation of amines and phenols in the lab?
2. Which of the following compounds can be estimated by quantitative acetylation?

    (a) Hydroquinone   (b) *p*-Anisidine

    (c) *N,N*-Dimethylaniline   (d) Isopropyl alcohol

3. Explain the role of blank titration in the estimation of aniline by quantitative acetylation.

4. Can quantitative acetylation of an amine/phenol be carried out in acetic anhydride alone?

5. Can the green acetylation be used for the quantitative estimation of aniline?

6. In the bromate-bromide method, estimation of aromatic amine or phenol is done by *in situ* generation of bromine. Explain why bromine cannot be used directly.

7. Besides, aromatic amines and phenols can you suggest any other compound which can be estimated by bromate-bromide method?

8. Estimation of aniline or phenol is done in the lab by using which type of iodine titration?

9. What is the role of concentrated hydrochloric acid in the bromate-bromide reaction?

10. Why acetic anhydride in pyridine is a better acetylating agent for phenols than acetic anhydride alone?

11. During estimation of phenol by bromate-bromide method, it is important to allow the reaction mixture to stand at room temperature for ten minutes after addition of KI. Explain.

12. Can you estimate sucrose by Fehling's solution method?

13. Why are Fehling A and Fehling B stored separately in the lab?

14. Explain why the Fehling's solution should be boiling during the estimation of glucose?

15. How can you improve the sensitivity of glucose estimation by Fehling's solution?

16. Why alcoholic and not aqueous potassium hydroxide is used in the estimation of the saponification value of an oil?

17. What is the need to titrate alcoholic KOH with oxalic acid in the saponification value determination? Can you use any other reagent in place of oxalic acid?

18. During determination of saponification value of an oil, it is important to convert volume of oxalic acid used, to volume of potassium hydroxide. Give reason.

19. Does transesterification affect the saponification value of an oil, during estimation?

20. Give commercial application of determining saponification value of an oil?

21. Can you predict the completion of saponification reaction by looking at the reaction medium?

22. What is the significance of determining iodine value of an oil?

23. Name two methods which can be used for estimation of amino acids.

24. In Sorensen's Formol titration, glycine and formalin solutions should be neutralized before mixing. Explain.

25. What will happen if you titrate the given solution of glycine with NaOH directly?

26. What is the approximate pK_1 value for a neutral amino acid? Do you expect pK_1 of acidic and basic amino acids to be the same or any different?

27. Do you expect the volume of NaOH required to neutralize a solution of acidic amino acid in water to be higher than that of neutral amino acid?

28. What is the need to titrate NaOH with oxalic acid, in Sorensen's method?

29. Can you use HCl in place of oxalic acid for standardization of NaOH?

30. What is the application of Sorensen's Formol titration, besides estimation of amino acids?

31. What is the advantage of modified Winkler's method over conventional one?

32. DO is an important water quality parameter. Explain why DO should not be too high or too low?

33. What precautions are required while collecting the water sample for DO measurement?

34. What is the role of KI in the Winkler's DO estimation?

35. Define BOD? How is it different from COD?

36. Why COD values are always higher than BOD?

37. A water stream has high BOD. Comment on the suitability of this stream for marine life.

38. What is the role of the following in COD determination?
    (a) Mercuric sulphate   (b) Silver sulphate
    (c) Ferrous ammonium sulphate

39. Can you estimate aspirin in a commercial sample by bromate-bromide method? If yes, how?

40. Estimation of aspirin by titrimetry is nothing but a simple acid-base titration. Explain.

41. Give one chemical test to prove the purity of an aspirin sample.

2

Extraction from Natural/Commercial Samples

2.1   INTRODUCTION

Extraction is an important technique used for the isolation and purification of organic compounds from natural sources. The interest in natural compounds stems from their numerous applications in fields like pharmaceuticals, cosmetics, dietary supplements, etc.

The extraction method involves preparing a solution of the source in one solvent and then mixing the second solvent that is immiscible with the first one. The desired solute is selectively extracted in the second solvent due to its higher solubility in the latter. Evaporation of the second solvent then gives the desired compound which can be purified and put to desired use.

The purpose of these experiments at undergraduate level is to introduce microscale extraction techniques and to allow the undergraduate students to practice this technique. These exercises demonstrate how extraction is used by organic chemists at different stages in industry and in research. Some of the representative extractions of natural compounds have been discussed in this chapter.

2.2   ISOLATION OF CASEIN FROM MILK

Basic Lab Operations: Precipitation.

Estimated Time: 45 minutes

Chemicals Required: Glacial acetic acid, ethanol, ether.

What You Should Know: Casein is a phosphoprotein which contains ~15 amino acids. It has phosphate groups attached to the hydroxyl groups of some of the amino acid side chains. It is an amorphous white coloured solid, insoluble in organic solvents.

Milk contains about 3% casein, wherein casein exists as a calcium salt, calcium caseinate. pH of milk (~6.7) is higher than the isoelectric point of casein (~4.6). So to precipitate casein, acetic acid is added drop by drop until the isoelectric point of the protein is reached. Casein in milk is present in the form of soluble anionic micelles. The white turbid appearance of milk is due to the presence of these micelles. Changing pH from 6.7 to 4.6 causes self-aggregation of neutral casein molecules leading to its precipitation.

Experimental Procedure: Place non-fat milk (100.0 mL), distilled water (300 mL) in a 500 mL beaker. Warm the contents on water bath to 40°C. Next, add 10% acetic acid dropwise, with constant stirring, till the precipitation of casein is complete. Filter the solid using a muslin cloth, wash with water and then ethanol. Place the casein so obtained in a beaker and add ether (ether will dissolve any fat present in milk), stir it for 10 minutes. Decant the solvent and dry the solid between the folds of filter paper. Leave it in the desiccator, overnight and weigh.

Note

- The isoelectric point (pI) of a protein is the pH at which it is electrically neutral *i.e.* net charge is zero. Protein exhibits lowest solubility at its pI.
- Supernatant liquid will become colourless when precipitation is complete.
- If excess CH_3COOH is added, the precipitated casein will redisssolve.

Result: The given sample of milk contains ------- g of casein.

Utility of This Exercise

- Casein is the chief constituent of cheese, is used in food, dietary supplements and in cosmetics.

Lessons to Take Home

After doing this exercise, the students will be able to do the following:

- Successfully precipitate casein from milk samples.
- Understand the role of pH in precipitation of proteins.
- Understand how concept of pI can be used for purification of a protein sample and for separation of a mixture of proteins.

2.3   ISOLATION OF EUGENOL FROM CLOVES

Basic Lab Operations: Steam distillation, solvent extraction.

Estimated Time: 1 Hour

Chemicals Required: Cloves, dichloromethane, potassium hydroxide (KOH), hydrochloric acid and sodium sulphate.

What You Should Know: Eugenol, the main constituent of essential oil, is obtained from *Syzygium aromaticum* (commonly known as clove). The concentration of eugenol in clove oil is as high as 90%. It is found in lower concentrations in bay leaves and spices as well as other botanical oils. Eugenol has a boiling point of 253.6°C and can be isolated from cloves at a lower temperature through steam distillation.

4-Allyl-2-methoxyphenol

Experimental Procedure: Weigh cloves (15.0 g) and crush them in a mortar and pestle, if the spice is not ground. Transfer the contents in a 250 mL round bottomed flask. And add water (80-100 mL). Soak the spice in water for 15 minutes. Set up a simple distillation apparatus, use water condenser. Swirl the flask gently from time to time. Continue the distillation process until 100 mL of distillate has been collected. Add water (50.0 mL) to the round bottomed flask after first 50.0 mL of distillate has been collected. Once the distillation process is over, transfer the distillate in a separatory funnel and extract with three portions (~15 mL each) of dichloromethane. Drain out the organic layer and transfer it into a clean separating funnel, and extract with 10% aqueous KOH solution. Eugenol, being a phenol, goes in the aqueous layer as potassium salt. Discard the organic layer and collect the aqueous layer. Acidify the aqueous layer with cold dilute hydrochloric acid. Again, extract the acidic solution with dichloromethane (15.0 mL) twice, drain the organic layer and dry it with anhydrous sodium sulphate. Allow the solution to stand over sodium sulphate for 10-15 minutes, with occasional stirring. Filter, the organic layer and distill in a rotatory evaporator. The pale-yellow oil obtained as residue is eugenol.

Characterization

- It has boiling point, 253.6°C.
- It gives a yellow green colour with ferric chloride (characteristic test for phenols)
- It forms brown–red coloured picrate on reaction with picric acid.

Result: The given sample contained ------- g of Eugenol, boiling point ------- °C (Literature Value 253.6°C).

Utility of This Exercise

- Eugenol is known to exhibit anti-fungal, anti-bacterial and insecticidal activities.
- It is also known to possess antioxidant, antiseptic and anti-inflammatory properties.
- It is added to insect attractant formulations for attracting fruit flies and melon flies.

- It is also used as an anaesthetic by hobby aquarists
- It is commonly used as a flavoring agent in food products, in cosmetics and particularly in dentistry.
- It is a constituent of mouth washes, tooth pastes, soaps, etc.

Lessons to Take Home

After doing this exercise, the students will be able to do the following:

- Extract eugenol from cloves.
- Understand the role of KOH in eugenol extraction.
- Understand how acid-base chemistry can be used for isolation and purification of organic compounds from a mixture.
- Extend this method to isolate/purify other acidic and basic compounds.

2.4 ISOLATION OF PIPERINE FROM BLACK PEPPER

Basic Lab Operations: Refluxing, solvent extraction.

Estimated Time: 24 hours

Chemicals Required: Ethanol, potassium hydroxide, acetone.

What You Should Know: Piperine is the alkaloid responsible for the pungency and biting taste of black pepper, *Piper nigrum* and *Piper longum L.*, commonly known as long pepper. Structurally, it is an amide which is present to the extent of 5-9% in black pepper. It is known to stimulate digestive enzyme of pancreas and is also used as a preservative. Extraction of piperine from black pepper can be conveniently carried out with 95% ethanol preferably using soxhlet apparatus. Alcoholic potassium hydroxide is added to the concentrated ethanolic extract of black pepper to keep the acidic resins in solution as soluble potassium salts and the clear solution is diluted with water and is kept overnight to precipitate the pure piperine crystals.

Piperine

or

(2*E*, 4*E*)-5-(benzo[*d*][1, 3]dioxol-5-yl)-1-(piperidin-1-yl)penta-2,4-dien-one

Experimental Procedure: Place ground pepper powder (10.0 g) and ethanol (95%, 120.0 mL) in a round bottomed flask fitted with a water condenser. (Use of soxhlet condenser is recommended!). Reflux the mixture for 3 hours. After cooling the flask, filter the pepper grounds. Transfer the filtrate into a round bottomed flask and concentrate it to ~10.0 mL using rotatory evaporator. Cool the residue in an ice bath, add warm alcoholic KOH solution (2 N,10.0 mL) to the flask, mix well and filter to obtain a clear filtrate. Warm the solution on water bath and add water (5-7 mL) dropwise. Turbidity is observed at this stage and yellow crystals may start separating. Keep the solution overnight and filter the yellow crystals of piperine. Recrystallize from acetone (use water bath). Filter the recrystallized solid, dry, record the yield and melting point of the solid.

Result: ------- g of piperine was obtained from the given sample of black pepper. Melting point of the recrystallized sample was found to be -------. (Literature melting point- 129-131°C).

Utility of This Exercise

- Piperine displays numerous biological effects such as anti-inflammatory, antioxidant, anti-mutagenic and antitumor activities.
- It is used in many herbal medicines and in food supplements to increase the bioavailability of other vitamins and minerals.

Lessons to Take Home

After doing this exercise, the students will be able to do the following:

- Extract pure piperine from black pepper.
- Extend this method for extraction of other useful compounds.

2.5 EXTRACTION OF CAFFEINE FROM TEA LEAVES

Basic Lab Operations: Refluxing, solvent extraction.

Estimated Time: 90 minutes

Chemicals Required: Black tea leaves, dichloromethane (DCM), sodium carbonate and sodium sulphate.

What You Should Know: Caffeine belongs to a very important class of compounds called **purine alkaloids**. Naturally, it is present in the fruit and bark of a number of plants including tea, coffee and cacoa. Tea bags contain about 30-75 mg and coffee bags 80-125 mg of caffeine in a typical 150.0 mL (cup) serving. Caffeine is a mild stimulant and a diuretic. It acts as a stimulant for central nervous system, skeletal muscles, and heart. It is also a diuretic (it promotes urination). Caffeine is highly addictive and is possibly the most widely abused drug in many countries. Symptoms of caffeine withdrawal may include headache, lethargy, insomnia or nausea.

Besides caffeine, tea leaves also contain cellulose, tannins, flavonoid pigments and chlorophylls (Figure 2.1). Tannins are phenolic compounds of high molecular weight.

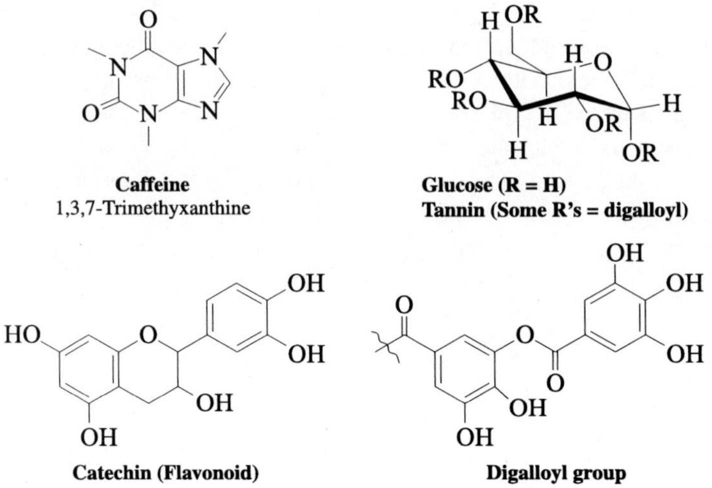

Caffeine
1,3,7-Trimethyxanthine

Glucose (R = H)
Tannin (Some R's = digalloyl)

Catechin (Flavonoid)

Digalloyl group

Fig. 2.1   Some Components of Tea Leaves

The differing solubility of these compounds forms the basis of extraction of caffeine from tea leaves. Caffeine is first extracted from tea leaves using hot water. The tannins, flavonoids, pigments and chlorophylls will also be extracted from the water, cellulose which is insoluble in water is removed at this stage. Since the flavonoids and tannins are acidic, base (sodium carbonate) is added to the aqueous extract in order to ionize them and ensure that they remain in the aqueous layer as soluble salts. Caffeine can then be extracted from the water extract using dichloromethane (DCM) as solvent. Chlorophyll, which is DCM soluble, also gets extracted along with caffeine. Still, you can get nearly pure caffeine after evaporating DCM layer on rotatory evaporator or on steam bath.

Experimental Procedure: Place six tea bags, distilled water (200 mL) along with sodium carbonate (6.0 g) in a 500 mL beaker. Cover the beaker with watch glass and boil the contents for 30 minutes. Allow the mixture to come to room temperature, transfer it to a separating funnel and extract with three portions (~15.0 mL each) of dichloromethane. Care should be taken not to mix the contents too vigorously as it may lead to formation of emulsion, which is difficult to break. After shaking for 3-4 minutes, allow the separating funnel to stand for 5-10 minutes until the two layers separate completely. Carefully collect the bottom organic layer and dry over crystalline and anhydrous sodium sulphate for 10-15 minutes. Repeat this procedure two more times and pool all the three portions of organic layer. Filter the organic layer and evaporate in a rotary evaporator (carefully) or on steam bath (**use fume hood**). Weigh the white crystalline solid so obtained and record its melting point.

Note

- Brown color of tea leaves is due to flavonoids, chlorophyll and their oxidation products.
- DCM is highly toxic and a suspected carcinogen, therefore the students should not excessively inhale its fumes and should avoid any kind of spill over clothes.
- Wear proper safety gear while handing DCM and carry out all operations in the fume hood.
- While extracting caffeine from water extract, release the pinch cock from time to time to prevent bursting of lid.
- In case emulsions are formed after adding DCM to water extract, try adding little bit of NaCl and shake gently to break the emulsion.
- Commercially, super critical carbon dioxide has replaced dichloromethane for extraction of caffeine. DCM, however can be used for small laboratory scale exercise.

Test for Caffeine

Warm a few milligrams of caffeine with potassium ferrocyanide and a few drops of nitric acid, a prussian blue colour is obtained.

Result: -------------- g of caffeine was obtained from the given sample of tea leaves. Melting point of the sample was found to be -------. (Literature melting point 238°C).

Utility of This Exercise

- The extraction of caffeine from coffee to get decaffeinated coffee is commercially important as it eliminates the harmful effects of caffeine on human health.
- It is used in proprietary drugs for stimulant effect to prevent drowsiness.
- It is also added to colas and chocolates for taste.

Lessons to Take Home

After doing this exercise, the students will be able to do the following:

- Understand how extraction of organic compounds can be undertaken from natural/commercial sources.
- Understand the role of base in caffeine extraction.
- Understand why methanol cannot be used in place of dichloromethane in this exercise.

2.6   EXTRACTION OF D-LIMONENE FROM ORANGE PEEL (CONVENTIONAL METHOD)

Basic Lab Operations: Steam distillation, solvent extraction.

Estimated Time: 1 hour

Chemicals Required: Nil

What You Should Know: Limonene, 1-methyl-4-prop-1-en-2-yl-cyclohexene, belongs to a class of cyclic terpenes and exist as a colourless oily liquid with the smell of oranges, at room temperature. Its molecular formula is $C_{10}H_{16}$ and boiling point is 176°C.

Limonene

D-Limonene, the active enantiomer, can be extracted from peel of oranges and other citrus fruits like lemon as well as pine trees. The extraction can be carried out either through steam distillation using water as a solvent or using liquid carbon dioxide.

In the conventional experiment described below, orange peels are boiled with water and the oil produced (limonene) is distilled in steam at a temperature < 100°C, well below normal boiling point of limonene. The immiscible oil is then separated by solvent extraction. Direct extraction by heating will result in decomposition of limonene, whereas steam distillation prevents this from happening.

Experimental Procedure: Grate the outer orange coloured rind of two oranges with a cheese grater. Transfer it to a round bottomed flask (250 mL) and add distilled water (100 mL) along with a few pumice stones. Set up the distillation apparatus and heat the flask so that distillation proceeds at a steady rate, approximately one drop per second of distillate. Stop distillation when clear distillate starts coming because as long as limonene is distilling, the distillate is slightly turbid. Collect approximately 50.0 mL of distillate and transfer it to a slender separating funnel, allow the oily layer of limonene to separate. Limonene forms the upper layer, being lighter than water. Drain out the lower aqueous layer. Transfer the remaining liquid in a test tube and carefully separate limonene with a dropper into another dry test tube. Dry the oily limonene in a dessicator.

Test for Limonene

Limonene is an unsaturated hydrocarbon which can be tested by using bromine water or alkaline potassium permanganate.

Result:................g of limonene was isolated from the given sample of orange peel.

Utility of This Exercise

- Limonene is a powerful antiinflammatory and antioxidant compound.
- It is used as a flavouring agent for food, beverages and chewing gums.
- It is also added to pharmaceutical creams and helps them to penetrate skin.
- It is an insecticide and a potential biofuel
- It is used as a fragrant and a cleansing solvent in wash liquids used to clean oil stains at airport runways and parking.

Lessons to Take Home

After doing this exercise, the students will be able to do the following:

- Successfully extract limonene from citrus fruits.
- Understand the role of steam distillation in separation and purification of organic compounds

2.7 EXTRACTION OF D-LIMONENE FROM ORANGE PEEL USING LIQUID CARBON DIOXIDE AS SOLVENT (GREEN METHOD)

Basic Lab Operations: Extraction

Estimated Time: 1 hour

Chemicals Required: Dry ice

What You Should Know: See section 2.6 for basic information on Limonene

Advantages of green method over conventional method

- The protocol involves use of polypropylene centrifuge tubes and liquid carbon dioxide.
- It is a fast and economical process which does not require costly apparatus.
- The protocol is easy to manoeuvre and workup.
- Product obtained is pure and lacks any side products or thermal degradation products.
- Dry ice is a green and environment friendly solvent. It can be easily removed from the extract due to low boiling point.

Experimental Procedure: Make a solid trap by bending copper wire into coils and a handle and place a filter paper on the loop. Now insert this solid trap in a 15.0 mL plastic centrifuge tube (See Figure 2.2). For extraction, introduce grated coloured part of the orange peel (~2.5 g) in the tube and fill the tube with crushed dry ice and seal with a cap. Thereafter, place the centrifuge tube in warm water (40-50°C) in the graduated cylinder. As the temperature rises, solid CO_2 liquefies and boils. It passes through the peel and moves to the bottom thus extracting the oil in it. As the heating continues, pressure begins to develop in the centrifuge tube and the carbon dioxide gas starts to escape from the tube slowly. Once the liquid has evaporated and evolution of gas has subsided, remove the tube from the cylinder and remove the cap. Collect the product from the tip of the tube as pale yellow liquid. Extract again with fresh sample of dry ice, if required.

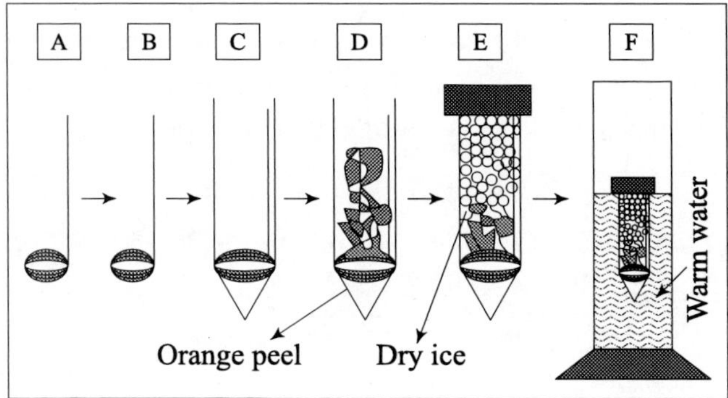

Fig. 2.2   Extraction of Limonene from Orange Peel Using Liquid Carbon Dioxide

Note

- One should wear gloves while dealing with dry ice as its direct contact may cause damage to the skin.

- Only plastic centrifuge tubes should be used for the experiment as the pressure generated during reaction may break it otherwise.

- Extraction can be repeated again in the same centrifuge tube with more dry ice, if required. But the same centrifuge tube should not be used more than five times for extraction.

Result: ------ g of limonene was obtained from the given sample of orange peel.

Utility of this Exercise: See experiment 2.6, page 43.

Lessons to Take Home

After doing this exercise, the students will be able to do the following:

- Understand why this method is superior to conventional method.
- Explain the role of dry ice in extraction of limonene.

2.8 ISOLATION AND ESTIMATION OF GENOMIC DNA FROM CAULIFLOWER

Basic Lab Operations: Extraction and spectrophotometry.

Estimated Time: 2 hours

Chemicals Required: Sodium dodecyl sulphate (SDS, 0.5g), ethylenediaminetetraacetic acid (EDTA, 0.0015 M, 0.0056 g in 10.0 mL water), sodium chloride (0.15 M, 0.088 g in 10.0 mL water), absolute ethanol (5.0 mL), trisodium citrate dihydrate (0.15 M, 0.0441 g in 10.0 mL water), chloroform: isopropyl alcohol (20:1, 0.5 mL), Tris-EDTA buffer (10.0 mL, see Appendix-II)

What You Should Know: Deoxyribonucleic acid (DNA) is the carrier of genetic information in almost all living organisms, except a few viruses. The primary function of DNA is to store and express genetic information in living beings. It plays vital role in the biological processes like replication, transcription, translation, gene repair, etc.

DNA can be extracted from almost any living or preserved tissue. Bulk of DNA in eukaryotes is located in the cell but small amounts are also present in mitochondria and chloroplast. Although the DNA in a cell is about 100,000 times as long as the cell is, it only takes up about 10% of the space inside the cell because it is highly folded and packed together with proteins into complex units called **chromosomes**.

Major problems encountered in isolation of pure and intact DNA are:
(a) Degradation by mechanical damage, (b) hydrolysis of DNA by nucleases, (c) contamination of DNA preparation with polysaccharides, proteins, RNA and other cellular components. However, the methods of isolation available minimize all these problems. This section discusses the method for isolation of genomic DNA from cauliflower.

The main steps involved in isolation of DNA from cell nucleus are:

1. *Homogenization or disruption of cells:* Grinding of plant material in mortar and pestle in the presence of isolation mixture (IM) helps in cell lysis and cleavage of cell wall and nuclear membrane. EDTA in the isolation mixture helps to pull out magnesium (II) or calcium (II) ions required for the stability of nuclear membrane and activity of DNase, which is a DNA degrading enzyme. SDS in the IM helps to disrupt the polar interactions in the cell wall and pulls out proteins and lipids from the cell membrane, leading to its disruption. Heat softens the phospholipid bilayer and helps penetration of the IM. It also helps to deactivate DNase, which could otherwise lead to hydrolysis of DNA.

Sodium citrate is a chelating agent, which causes cell debris (degraded cell membrane, polysaccharides, proteins, lipids) to precipitate.

2. *Dissociation of nucelo-protein complex:* Detergents like SDS in the IM also help in disrupting DNA-proteins complex and degrading proteins while keeping DNA intact. Alkaline pH and high salt concentration help in this process. Cooling slows down the DNA breakdown by inactivating the DNase. Centrifugation settles the cell debris while DNA and proteins go into solution.

3. *Removal of contaminating material and precipitation of nucleic acids:* Proteins are denatured and precipitated by treating of the centrifugate with chloroform-isoamyl alcohol (20:1). Upon centrifugation, proteins form the middle layer between the aqueous (top layer) and organic layer (bottom layer). Other contaminants remain in the organic layer while DNA is still present in the aqueous layer. Small amount of NaCl added will bind to negatively charged phosphate groups of DNA whereby neutralizing it and helping its strands to coalesce and come together.

4. *Precipitation of DNA:* Addition of ice-cold ethanol reduces the solubility of polar sugar and phosphate groups and causes DNA to precipitate. Centrifugation now leads to pure DNA pallet. This pallet is dissolved in Tris-EDTA buffer(TE) and purity of DNA sample can be estimated by calculating the A_{260}/A_{280} ratio.

As per rough hypothesis, if the A_{260}/A_{280} ratio is between 1.7-2.0, then DNA sample is pure. Low A_{260}/A_{280} ratio indicates presence of proteins, which absorb strongly around 280 nm.

Note: Nucleotide bases of DNA absorb at 260 nm whereas aromatic amino acids in proteins absorb at 280 nm.

- Figures 2.3(a)–2.3(d) depict the isolation process of DNA from a plant cell.

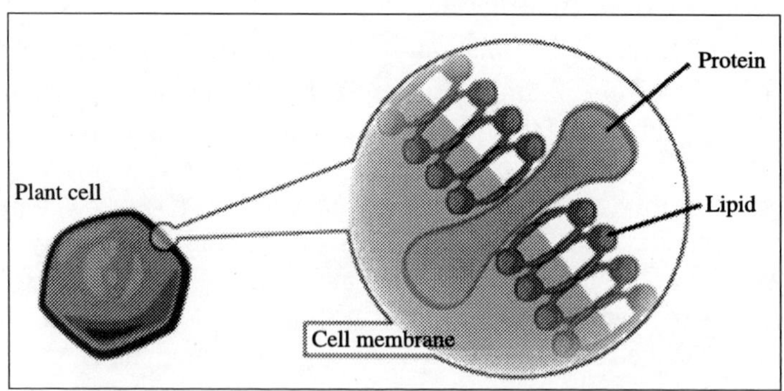

Fig. 2.3(a) Plant Cell Membrane having Lipoprotein Bilayer

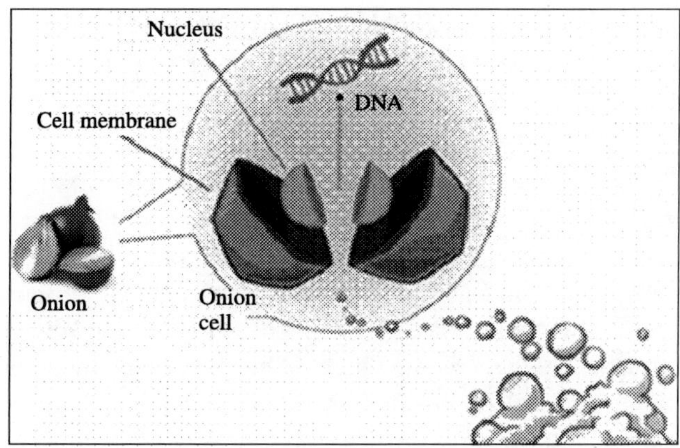

Fig. 2.3(b) Cell Lysis-Opens Cell Membrane and Nuclear Membrane

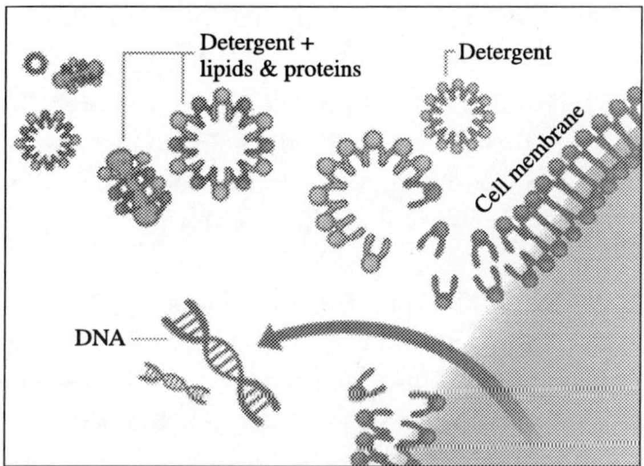

Fig. 2.3(c) Opening of Cell Membrane through pulling out of Lipids and Proteins by SDS

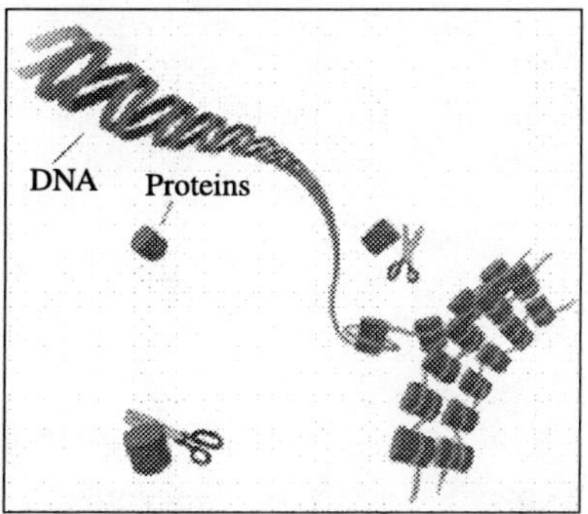

Fig. 2.3(d) Deproteinization-Pulls out Protein from DNA–Protein complex so DNA strand is easily workable

Experimental Procedure

- Prepare isolation mixture (IM, 10.0 mL): Take a   clean and dry beaker (100 mL). Pipette NaCl solution (7.5 mL), sodium citrate solution (1.0 mL), EDTA solution (1.0 mL) and add SDS (0.5 g). Heat the beaker to 60°C in water bath to dissolve SDS.

- Rinse the mortar and pestle with the isolation mixture. This avoids drying of isolation mixture in dry mortar and pestle. Take grated cauliflower (2.0 g, kept at –70°C for 3-4 hours, add isolation mixture (4.0 mL) and grind gently in mortar pestle for 2-3 minutes. This step is crucial, the quality of DNA depends on grinding, better the grinding better will be the DNA. There should not be any frothing during grinding, this can be ensured by grinding in one direction.

- Transfer the above mixture (the liquid) in 1.5 mL eppendorf and heat in a water bath maintained at 60°C for 10 minutes. Ready a beaker with crushed ice.

- Put the eppendorf in between the ice for 5 minutes.

- Centrifuge the contents at 10,000 rpm for 10 minutes.

- Prepare a mixture of chloroform and isoamyl alcohol (20:1). Transfer the supernatant liquid (1.0 mL), with the help of an autopipette, into another eppendorf and add organic solution (chloroform: isoamyl alcohol, 1.0 mL) and a pinch of sodium chloride. Shake gently twice or thrice.

- Centrifuge at 10,000 rpm for 30 minutes. After formation of three distinct layers-the bottom layer, the interface and the top layer has taken place, transfer the top layer gently and carefully (without disturbing the interface) into another eppendorf using autopipette.

- Slowly add chilled ethanol (–20°C) to aqueous solution, gently shake and watch the fibres of DNA separate out.

- *Estimation:* Centrifuge the mixture at 10,000 rpm for 10 minutes, take out the pellet and dry it completely (on smelling there should not be any smell of ethanol) and then dissolve the pellet in Tris-EDTA buffer (2.0 mL). Transfer this solution (0.4 mL) to a fresh eppendrof and dilute with water (3.6 mL). Transfer this solution to a cuvette and note the absorbance value at 260 nm and 280 nm (set the instrument sequentially at 260 and 280 nm, set the zero with distilled water on both occasions). Calculate the A_{260}/A_{280} ratio using UV spectrophotometer.

Note

- All glassware should be sterilized or autoclaved.
- Wear gloves to avoid contact of experimental material and apparatus with nucleases found in skin exudates.

- Keep absolute ethanol overnight at $-20°C$.
- Put chloroform and isoamyl alcohol in freezer. Take out from freezer during preparation.
- The three layers formed after centrifugation for 30 minutes are:
  (a) Bottom layer (organic): contains cell debris and other material.
  (b) Central layer: Proteins
  (c) Top layer (aqueous): contains DNA in solution
- A_{320} eliminates errors due to contaminants
- For preparation of Tris-EDTA buffer, see Appendix-II.

Observation and Calculations:

Concentration of DNA (µg/mL): $(A_{260} - A_{320}) \times$ dilution factor $\times$ 50

It is assumed that $A_{260} = 1$ for 50 µg/mL of DNA

Dilution factor = 10 (0.4 mL diluted to 4.0 mL)

$\therefore$ *DNA yield* (µg) = Concentration of DNA (µg/mL) $\times$ total sample volume(mL)

Result: ------µg of DNA was isolated from the given sample of cauliflower. A_{260}/A_{280} ratio for the isolated sample was found to be -------.

Utility of This Exercise

DNA isolation is useful for many purposes including:

- Diagnosis of medical conditions, to genetically engineer plants and animal and for drug development.
- Forensic science: To gather evidence in crime invstigations.
- Genome sequencing, detecting bacteria and viruses in the environment and determining paternity.

Lessons to Take Home

After doing this experiment, the students will be able to do the following:

- Isolate genomic DNA from other plant sources.
- Establish the purity of DNA sample.
- Understand the role isolation mixture, chloroform and isoamyl alcohol mixture and cold ethanol in DNA isolation.
- Understand why sodium chloride is important for DNA isolation and precipitation.

2.9 ISOLATION AND ESTIMATION OF GENOMIC DNA FROM ONION PEEL

Basic Lab Operations: Extraction and spectrophotometry.

Estimated time: 2 hours

Chemicals required: Sodium lauryl sulfate (SLS) (0.17M, 12.25 g), EDTA (Ethylenediaminetetra acetic acid, 0.0007M, 0.065 g), sodium chloride (0.15M, 2.19 g), ethanol (absolute, 50.0 mL), sodium acetate anhydrous (0.05M, 1.03 g), eye lens solution, double distilled water, peeled and grated onion.

What You Should Know: Onion is regarded as an ideal material for isolating DNA because of the following reasons:

- It has low starch content, *i.e.*, it does not have a lot of cellulose fibers, which allows the DNA to be more clearly seen.
- DNA abundance: enough DNA can be extracted and can be seen with the unaided/naked eye.
- Low cost

For detailed theory on DNA isolation, see *section 2.8.*

Experimental Procedure

1. *Prepare Cell Lysis Solution (or Homogenizing medium):* Mix Sodium lauryl sulfate (SLS/SDS, 12.25 g), sodium chloride (2.19 g), sodium acetate (1.03 g) and EDTA (0.065 g) in a volumetric flask (250 mL) and add distilled water to make up the volume to 250 mL.

2. *Homogenate*: Peel and grate a medium sized onion (~50.0 g) to prepare a homogenate. Mix the homogenizing medium (lysis solution) and onion homogenate in a 250 mL beaker.

3. *Heat the preparation*: Heat the beaker (containing the homogenate suspended in homogenizing medium) over the hot plate/water bath at 60-70°C for about 15 minutes with continuous stirring.

4. *Cool the preparation*: Quickly cool the preparation by placing it on ice. Keep stirring the solution gently to allow even cooling throughout.

5. *Filter the preparation*: Filter the homogenate through cheese cloth or filter paper.

6. *Deproteinize the homogenate*: Add 4-6 drops of eye lens cleaning solution to the filtered homogenate and keep it aside for 5 minutes. This step is actually DNA purification which can be excluded if students are not doing the DNA estimation or if they are not using DNA further for any other experiment.

7. *Precipitation of DNA*: Pour the clear solution or the filtered homogenate on a wide Petri dish. Add ice-cold absolute ethanol (50.0 mL) slowly along the side of the Petri dish. A clear layer of ethanol should form on top of the onion filtrate in Petri dish. Keep the solution undisturbed for 3-5 minutes. Bubbles will form precipitating the DNA, which becomes visible as the white strings/threads in the upper layer of ethanol.

8. *Spooling*: Spool out the stringy DNA onto a glass rod by slowly rotating the glass rod in one direction. Continue to rotate the glass rod as you

move it in large circles through the petridish, until complete DNA settles on it. Positive charges on silica glass attract negatively charged DNA stand.

9. *Washing and storage of DNA*: Transfer the spooled DNA from glass rod to a microcentrifuge tube/Eppendorf (2.0 mL) and add sufficient amount of absolute ethanol to the tube. Invert the tube gently a few times to settle down the DNA strands. Centrifuge the tube for 5 min and decant the alcohol. Repeat the process 4-5 times using 70% aqueous ethanol. The washed DNA can be stored at 4°C in 70% ethanol for further experiments or DNA estimation.

10. *Estimation:* Follow the procedure from *section 2.8*.

Observation and Calculations:

Concentration of DNA (μg/mL): $(A_{260}-A_{320})$ × dilution factor × 50

It is assumed that $A_{260} = 1$ for 50 μg/mL of DNA

Dilution factor = 10 (0.4 mL diluted to 4.0 mL)

DNA yield(μg) = Concentration of DNA (μg/mL) × total sample volume (mL)

Note

- The onion pieces should be of medium size, neither big nor small.
- Heating of the preparation containing the homogenate suspended in homogenizing medium should be done for 15 minutes with constant stirring.
- After boiling, cool the solution immediately. Stir vigorously while cooling.
- Absolute alcohol used for washing should be ice-cold.
- Spooling should be done by rotating the glass rod in one direction only. This will prevent DNA strands from breaking.
- For other points, see Note for *section 2.8*

Result: ------μg of DNA was isolated from the given quantity of onion peel. A_{260}/A_{280} ratio for the isolated sample was found to be -------.

Utility of This Exercise: See *section 2.8*

Lessons to Take Home: See *section 2.8*

Post Lab Questions

1. Explain the role of isoelectric point in the precipitation of casein from milk.
2. Which of the following compounds can be identified by simple chemical tests in the lab (like unsaturation, functional group test etc)? Name those tests.

   (a) Eugenol   (b) Limonene   (c) Caffeine

3. Sometimes during extraction of organic compounds with organic solvents, emulsions are formed, which render the solution inseparable. Suggest some methods to solve this problem.

4. Eugenol gives decolorization with both potassium permanganate and bromine water. Explain.

5. Piperine is hydrolysed with alkali to form a brown colored liquid and aqueous solution on acidification gives a crystalline solid, which gives brisk effervescence with sodium bicarbonate. Explain.

6. What is the role of base in the extraction of caffeine from tea leaves?

7. Steam distillation is a better method than simple distillation for extraction of limonene from orange peel. Give reason.

8. What is super critical carbon dioxide? List the uses of super critical fluids as solvents for extraction.

9. DNA is the only carrier of heredity. Justify this statement.

10. List the roles of DNA in a biological system.

11. Write the names of the bases and sugar found in DNA.

12. Cell lysis leads to which of the following:

    (a) Opening of cell wall

    (b) Opening of cell membrane

    (c) Breakdown of DNA-Protein Complex

13. In which ionic form does DNA exist at the physiological pH?

14. Why onion is a better source for DNA isolation?

15. List the major problems in isolation of DNA from cauliflower and explain how the learnt method addresses them.

16. What are the constituents of isolation mixture (IM) used in DNA isolation?

17. Why it is important to rinse the mortar pestle with IM before grinding cauliflower?

18. What is the role of the following in DNA isolation:

    (a) Grinding in presence of IM   (b) SDS

    (c) High NaCl concentration   (d) Alkaline pH

    (e) Heating after Grinding   (f) EDTA

    (g) Sodium citrate   (h) Cooling after heating

    (i) $CHCl_3$: Isoamyl alcohol   (j) Cold ethanol

19. During DNA isolation, why is NaCl added after addition of $CHCl_3$+ Isoamyl alcohol?

20. How will you know whether the obtained sample of DNA is pure?

21. What is the role of TE-Buffer in the DNA isolation? Can you use acetate buffer in place of TE buffer?

22. Why glass rod is used for spooling DNA?

3

Biochemical Estimation

3.1 INTRODUCTION

Biochemistry is the study of "chemistry of life". It helps the chemists to get basic understanding on how food products are converted into energy or useful chemical substances, how enzymes (biological catalysts) catalyse biological reactions and how they differ in catalytic ability, specificity and regulation from other organic/inorganic catalysts. It explains how DNA acts as carrier of heredity, how genes are transferred from one generation to another and how do drugs, pesticides and pollutants affect life. The study of biochemistry also helps us in creation of safer synthetic drugs by understanding their mechanisms of action, receptor-ligand binding interactions, it helps forensic scientists to solve crimes, allows development of safer agricultural and food products and much more.

This chapter deals with the basic biochemistry experiments prescribed by latest UGC syllabus including the CBCS/LOCF format for the undergraduate chemistry students.

3.2 STUDY OF THE ACID-BASE TITRATION CURVE OF GLYCINE AND DETERMINATION OF ITS ISOELECTRIC POINT

Basic Lab Operations: pH meter titration.

Estimated Time: 120 minutes

Chemicals Required: Oxalic acid, hydrochloric acid, sodium hydroxide, phenolphthalein and glycine solution.

What You Should Know: Acid-base titration curves are plotted by studying the variation in pH for a given volume of sample solution, on successive addition of acid or alkali. The curves are usually plots of pH of analyte against the volume of titrant added (see Figure 3.1). The aim of such titrations is to accurately locate

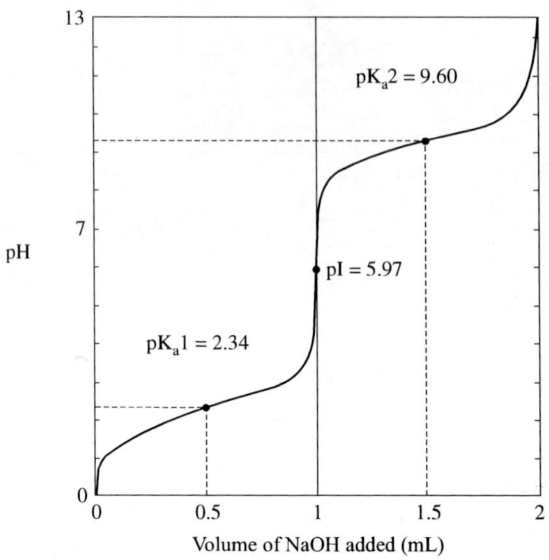

Fig. 3.1   Titration Curve of Glycine

the end point of the "S" shaped acid-base titration curve. This curve empirically defines several characteristics. The precise number of each characteristic depends on the nature of the specie being titrated. Broadly, they can be: (1) the number of ionizing groups, (2) the pKa of the ionizing group(s), (3) the buffer region(s), (4) Isoelectric point (pI) in case of an amino acid and (5) Molecular mass of the unknown amino acid.

The end point of the acid-base titration lies at the centre of the S-shaped curve, *i.e.*, the point at which there is steepest change of slope. Determining inflection point from a slanting S-shaped curve is however, a tedious job. A better method therefore is to plot the first derivative of the titration curve. The first derivative plot is easily produced by calculating differences between adjacent pH values and titrant volumes (V), taking ratios (ΔpH/ΔV) and plotting it against average volume, V_{av}. The inflection point in the curve is then identified as the point of maximum for an acid (or minimum for a base) as shown in Figure 3.2.

Amino acids are amphoteric substances which contain an acidic group (–COOH) as well as a basic group (NH$_2$) attached to the α-carbon. Depending upon the pH, these amino acids are capable of existing in one of the three forms, *i.e.*, conjugate acid (CA), dipolar ion (DI) and conjugate base (CB). Simple diprotic amino acids exist predominantly in the dipolar ion form in water. When the aqueous solution of amino acid is titrated with acid, it behaves like a base (*accepting proton to convert COO$^-$ to COOH*) and with base, it

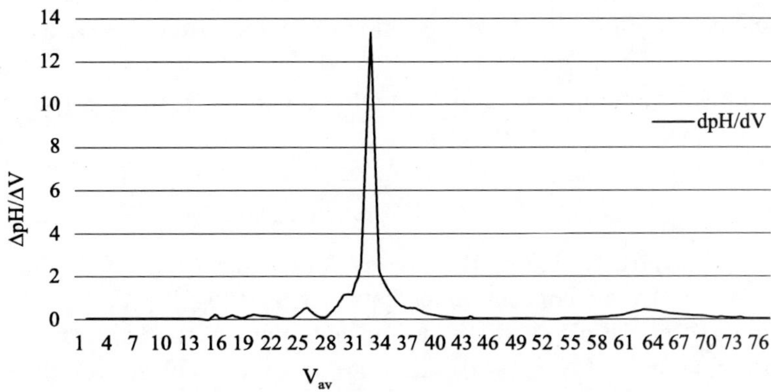

Fig. 3.2 First Derivative Curve for Variation of $\Delta pH/\Delta V$ with Average Volume (V_{av})

behaves like an acid (*losing proton from* NH_3^+). Thus, amino acids can be subjected to acid-base titration just like any other diprotic acid. Certain amino acids also have ionisable groups in the side chain which contribute additional ionisable groups to the molecule.

$$\underset{\text{Conjugate Acid}}{\overset{H}{\underset{R}{\overset{|}{\underset{|}{H_3\overset{\oplus}{N}-\!\!-COOH}}}}} \xrightarrow[\text{of base}]{\text{Addition}} \underset{\text{Zwitter ion}}{\overset{H}{\underset{R}{\overset{|}{\underset{|}{H_3\overset{\oplus}{N}-\!\!-COO^{\ominus}}}}}} \xrightarrow[\text{of base}]{\text{Addition}} \underset{\text{Conjugate Base}}{\overset{H}{\underset{R}{\overset{|}{\underset{|}{H_2N-\!\!-COO^{\ominus}}}}}}$$

$pH < pK_{aCOOH} < pK_{aAmmonium\,ion}$ $\qquad$ $pK_{aCOOH} < pH < pK_{aAmmonium\,ion}$ $\qquad$ $pK_{aCOOH} < pK_{aAmmonium\,ion} < pH$

Let us consider the case of glycine which is a diprotic amino acid with two dissociable protons in the acidic medium, one on the α-amino group and the other on the carboxyl group. Note that in the case of glycine, the side chain, R, does not contribute a dissociable proton.

Dissociation 1

$$\overset{H}{\underset{R}{\overset{|}{\underset{|}{H_3\overset{\oplus}{N}-\!\!-COOH}}}} \xrightarrow{\;pKa_1\;} \underset{\text{Zwitter ion}}{\overset{H}{\underset{R}{\overset{|}{\underset{|}{H_3\overset{\oplus}{N}-\!\!-COO^{\ominus}}}}}} + H^{\oplus}$$

Dissociation 2

$$\underset{\text{Zwitter ion}}{\overset{H}{\underset{R}{\overset{|}{\underset{|}{H_3\overset{\oplus}{N}-\!\!-COO^{\ominus}}}}}} \xrightarrow{\;pKa_2\;} \overset{H}{\underset{R}{\overset{|}{\underset{|}{H_2N-\!\!-COO^{\ominus}}}}} + H^{\oplus}$$

The dissociation of glycine, upon addition of base, depends on the acidity of the proton. The more acidic group will dissociate first (*i.e.*, at lower pH).

So, the proton on the α-COOH group will dissociate first (pH at this point will be called pK_1) followed by dissociation of proton of the $-NH_3^+$ group (pH at this point will be called pK_2).

The ionization of amino acid in solution follows the Henderson Hasselbalch equation:

$$pH = pK_a + \text{Log} \frac{[\text{Unprotonated form (base)}]}{[\text{Protonated form (acid)}]}$$

When the concentration of the ionised form equals that of the unionised form, the ratio of their concentrations equals 1, and log 1 = 0. Therefore, we can define **pK_a** as the pH at which the concentrations of the unprotonated form (base) and protonated forms (acid) of amino acid are equal. It should be noted that amino acid is at its best buffering capacity when its pH equals that of pK_a (that is, the solution resists changes in pH most effectively at this pH). The pK_a can be located on the S-shaped acid-base titration curve at midpoint of the buffering region, *i.e.,* the region where the pH remains practically constant on addition of the titrant (See Figure 3.1).

Each amino acid in solution is characterised by a unique **isoelectric point**, pI. Isoelectric point of an amino acid is the pH at which concentration of the dipolar ion is maximum or the pH at which the negatively charged (CB) species exactly equals the positively charged (CA) species. Knowledge of pI is useful in separation and purification of mixtures of amino acids and proteins. For diprotic (neutral) amino acid, the pI is the average of **pK_1** and **pK_2** values. For acidic amino acids, the pI is the average of pK_1 and pK_2 and for basic amino acids, it is the average of pK_2 and pK_3.

(**Note:** Acidic and basic amino acids have three dissociation constants due to ionisable side chains).

In this experiment, a diprotic amino acid of particular mass is dissolved in acidified water. Sodium hydroxide solution is then added dropwise to this solution and variation in pH is noted. By plotting the S-shaped graph and first derivative graph, you will be able to identify dissociation constants (pK_1 and pK_2), pI as well as number of ionizable groups for the amino acid under study. This method can also be used for identification of unknown amino acid by quantification of its molecular mass.

Experimental Procedure

1. Prepare ~ 0.5 N NaOH and ~ 0.5 N HCl solutions. Standardize the solutions as per the procedure give in Note, page 58.

2. Prepare standard glycine solution (100 mL, 0.5 N) by dissolving suitable amount of glycine in aqueous HCl (0.5 N). (See calculations)

3. Calibrate the pH meter with standard buffer solutions of pH 4.0, 7.0 and 9.0 respectively.

4. Transfer standard glycine solution (20.0 mL) into a 100 mL beaker with the help of a burette (Do Not Pipette!).

5. Add a glass rod in the beaker to enable mixing of the solutions upon addition.

6. Dip the probe of pH meter into the beaker such that it is fully immersed in glycine solution but does not touch the bottom/sides of the beaker or the stirrer of the beaker. Note the initial pH of the solution using pH meter (it should be ~1.5).

7. Take standardized 0.5 N NaOH in the burette. Do not forget to rinse the burette with 0.5 N NaOH solution before transferring it to the burette.

8. Begin titrating the glycine solution with the standardized sodium hydroxide solution by adding 0.5 mL aliquots from the burette. Stir the solution after every addition, record the volume of NaOH being added and also note down the pH.

9. When pH of the solution starts to change more rapidly (pH ~ 3.0), add 0.2 mL aliquots of NaOH solution till the pH reaches around 11.0.

10. Now again start adding 0.5 mL aliquots of NaOH solution till the pH reaches 12.5.

11. Tabulate your results and calculate ΔpH, ΔV and V_{av}, (see Table 3.1).

12. Now plot the S-shaped graph of pH v/s volume of NaOH added and also plot the first derivative graph by plotting ΔpH/ΔV v/s V_{av}.

13. Determine the pK_1, pK_2 and pI from the first derivative and the S-shaped graphs.

Observations and Calculations

Table 3.1 Titration of Glycine v/s NaOH

S. No.	Volume of NaOH added (mL)	pH	ΔpH	ΔV	ΔpH/ΔV	V_{av} (mL)

Normality of Standardized HCl = N

∴ Amount of glycine required to prepare 100 mL of N normal solution:

$$= \frac{75 \times 100 \times N}{1000}$$

$$= \mathbf{x\ g}$$

In the first derivative graph:

- Volume of NaOH corresponding to first inflection point = Volume of NaOH required for complete ionization of CA($^+$NH$_3$CH$_2$COOH) = V_1
- Volume of NaOH corresponding to second inflection point = Volume of NaOH required for complete ionization of DI ($^+$NH$_3$CH$_2$COO$^-$) = V_2

$$\therefore \qquad pK_1 = pH \text{ at } \frac{V_1}{2}$$

Similarly, $$pK_2 = pH \text{ at } \left(V_1 + \frac{V_2 - V_1}{2} \right)$$

$$\therefore \qquad pI = \frac{(pK_1 + pK_2)}{2}$$

Note

- If the concentration of the acid is known (number of moles) then we can manually calculate the number of moles of the base required for complete neutralization of the acid. The volume of the base used will thus correspond to the end point.
- Don't pipette NaOH or HCl/acidic glycine solution.
- Addition of NaOH to glycine solution should be done slowly so that accurate measurement of points where pH changes sharply can be done.
- The standardization of NaOH and HCl solution (0.5 N) can be done as follows:

    (i) Prepare standard oxalic acid (0.5 N). Titrate with ~0.5 N NaOH and determine its normality.

    (ii) Use the standardised NaOH to determine the normality of ~0.5 N HCl solution.

    (iii) Use phenolphthalein as an indicator in both cases.

- The normality of glycine solution should exactly equal normality of standardized HCl solution.
- pK_1 is the pH corresponding to half neutralization of CA form. Use the 'S' shaped curve to determine pK_1 at $\dfrac{V_1}{2}$.
- pK_2 is the pH corresponding to half neutralization of DI form. Use the 'S' shaped curve to determine pK_2 at $\left(V_1 + \dfrac{V_2 - V_1}{2} \right)$

Result: The pK_1, pK_2 and pI for the given amino acids were found to be ------, -------- and ------ respectively.

Utility of This Exercise

This experiment can be used for the following:

- Determining the number of ionisable groups in an amino acid.
- Determining the pK_a values of amino acids
- Determining the pI values of amino acids and thus categorising them as acidic, basic or neutral.
- Determining molecular weight of unknown amino acids.

Lessons to Take Home

By doing this experiment, the students will be able to do the following:

- Understand the acid-base behaviour of amino acids
- Predict the shapes of the titration curves for acidic, basic and neutral amino acids.
- Understand the concepts of buffering region, inflection point, dissociation constants and isoelectric point.
- Determine the molecular mass of the unknown amino acid from the data.
- Identify the unknown amino acid from its pK_a, pI values, and molecular mass.

3.3 PROTEIN ESTIMATION

Proteins are an important class of **biomolecules**. The word 'protein' was coined by Mudler (1938) and finds its origin in the Greek word 'proteos' which means 'first'. First, because proteins are the most important substances of life which are responsible for holding the animal life together and running it. They are found in cytoplasm and cell membrane of all living cells, but are principal components of skin, tendons, muscles, nerves, blood, enzymes, antibodies, hormones, etc. They are also present in articles of daily use like silk, wool, leather, etc.

Naturally occurring proteins are polyamides composed of ~20 α-amino acids, with exception of proline and hydroxy proline which are α-imino acids. These polyamides having molecular weight more than 10,000 daltons are called **proteins**. If, however molecular mass is less than 10,000 daltons, they are referred to as **polypeptides**. The amino acids in protein are joined together by an amide bond (called **peptide bond**) formed between the carboxyl and amino groups of the adjacent amino acid residues (Figure 3.3).

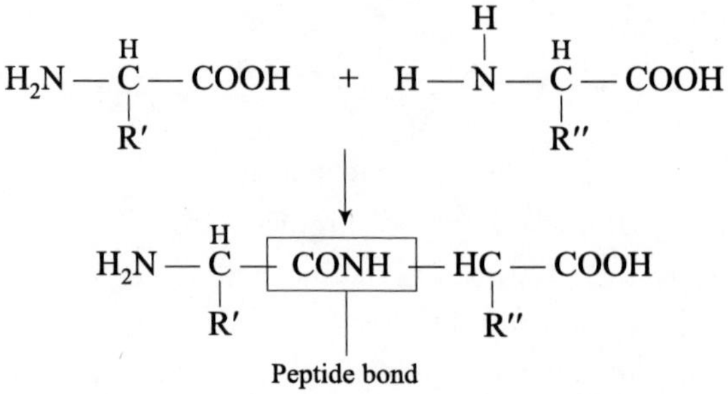

$$H_2N - \underset{\underset{R'}{|}}{\overset{\overset{H}{|}}{C}} - COOH \quad + \quad H - \underset{\underset{R''}{|}}{\overset{\overset{H}{|}}{N}} - \underset{}{\overset{\overset{H}{|}}{C}} - COOH$$

$$H_2N - \underset{\underset{R'}{|}}{\overset{\overset{H}{|}}{C}} - \boxed{CONH} - HC \underset{\underset{R''}{|}}{} - COOH$$

Peptide bond

Fig. 3.3 Formation of Peptide Bond

Proteins perform various useful functions in the body. They act as catalysts (enzymes), help in regulation of biological processes (hormones), in growth (building new cells), in maintenance (replacement of proteins lost by wear and tear of tissues), in defence (antibodies) and in communication (nerves), etc.

Methods of Protein Estimation: Depending on the accuracy required, the amount and purity of the protein available, compatibility of measurable sample with the reagents and time required for assay, various methods are appropriate for determining protein concentration in biological samples. Most popular methods of protein estimation are based on colorimetric assays where a protein is made to combine with a metal/dye to give a complex, which absorbs in the UV-visible range of the spectrum. The common methods of protein estimation are listed below. Each of these methods has some advantages and also suffers from some limitations. Amongst these methods, Lowry method is most useful for an undergraduate lab, being sensitive, low cost and reasonably fast. The methods are:

1. Micro Kjeldahl method
2. Lowry method
3. Biuret method
4. Bradford method
5. Spectrophotometric method
6. Turbidimetric method
7. BCA method

Role of Buffers in Biochemical Processes: A buffer is a solution which resists change in the pH on addition of small amount of acid or alkali. Ideally, buffers work best in the pH range of +/- 1. Buffers are extensively used in the biochemical reactions, since they aid in maintaining the pH of the medium while performing the various laboratory operations like extraction, isolation and the purification of various biomolecules. Selection of appropriate buffer with optimal pH is important as it may have a profound influence on extractability, stability and even biological functioning of the cell constituents. For example, all biochemical reactions are catalysed by the enzymes whose stability and activity depends on the pH of the system. In many of these biological reactions, protons may be consumed or released. In all such reactions, it is important to provide a system, *i.e.*, buffer, which could stabilize the H^+ ion concentration and obviate adverse effects due to changes in pH, on the enzymes under investigation.

Criterion for the Selection of the Buffer: While choosing a buffer, the following factors need to be taken into account. It should:

1. Possess adequate buffering capacity in the required pH range.
2. Be chemically inert, not bind and react with other molecules. This especially holds for assaying the activity of the enzyme as we need to add metal ions as activators/inhibitors.
3. Be available in high degree of purity and not contain the impurities which interfere with the estimations.
4. Be enzymically and hydrolytically stable.
5. Not be influenced by the changes in temperature, ionic composition and concentration or the salt effect of the medium.
6. Be non-toxic.
7. Not absorb light in the UV-Visible region.

Some of the commonly used buffers in the protein chemistry are:

1. Acetate buffer (acetic acid/sodium acetate, pH range: 3.6-5.6)
2. Phosphate buffer (NaH_2PO_4/Na_2HPO_4, pH range: 5.8-8.0)
3. Citrate-phosphate buffer (citric acid/Na_2HPO_4, pH range: 2.6.-7.6)
4. Tris buffer (pH range: 7.2-9.0)

3.3.1 Estimation of Amount of Protein in the Given Sample by Lowry Method

Basic Lab Operations: Spectrophotometry.

Estimated Time: 120 minutes

Chemicals Required: Bovine Serum Albumin (BSA) (200 µg/mL solution in distilled water); analytical reagent (0.5 mL of 1% aqueous $CuSO_4$, 0.5 mL of 2% aqueous Rochelle salt, 49.0 mL of 2% Na_2CO_3 in 0.1 M NaOH); Folin-Ciocalteu reagent [phosphomolybdic acid and phosphotungstic acid in phenol (1.0 N), dilute the commercial reagent, 2.0 N, 1:1 with distilled water, just before addition].

What You Should Know: Lowry method is the most commonly used method for estimation of proteins in cell free extracts because of its high sensitivity and also because quantities as low as 5 µg/mL of proteins can be measured.

The peptide bonds (-CO-NH-) in polypeptide chain reacts with copper sulphate in alkaline medium to give tetrahedral blue colored complex. The tyrosine and tryptophan residues of protein cause reduction of phosphomolybdate and phosphotungstate components of **Folin-Ciocalteu** reagents to give bluish product which contributes towards enhancing the sensitivity of this method. It is however important to remember that several compounds like EDTA, carbohydrates, Tris, NH_4^+, K^+, Mg^{2+} ions, thiol reagents, phenols, etc., interfere with color development and it should be ensured that such substances are not present in sample preparation.

The Folin-Ciocalteu reagent (FCR) or Folin's phenol reagent or Folin-Denis reagent, is a mixture of phosphomolybdic and phosphotungstic acid used for the colorimetric assay of phenolic and polyphenolic antioxidants. It works by measuring the amount of the substance, being tested, needed to inhibit the oxidation of the reagent. However, this reagent does not only measure total phenols but reacts with any reducing substance, compound having hetero atoms and conjugated systems. The reagent therefore measures the total reducing capacity of a sample, not just the level of phenolic compounds. This reagent forms part of the Lowry protein assay and also reacts with some nitrogen-containing compounds such as hydroxylamine and guanidine.

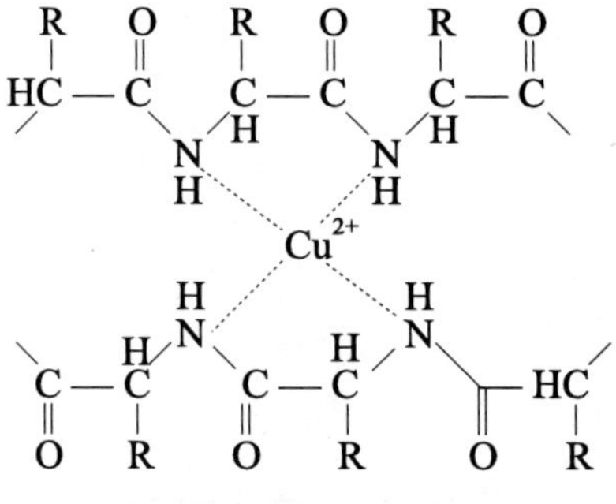

Tetradentate Copper complex

$$\text{Reducing groups in the peptide/amino acids} + Mo^{6+}/W^{6+} \xrightarrow{\substack{\text{Alkaline} \\ \text{medium}}} \text{Blue complex absorbs at 660 nm} \ (Mo^{5+}/W^{5+})$$

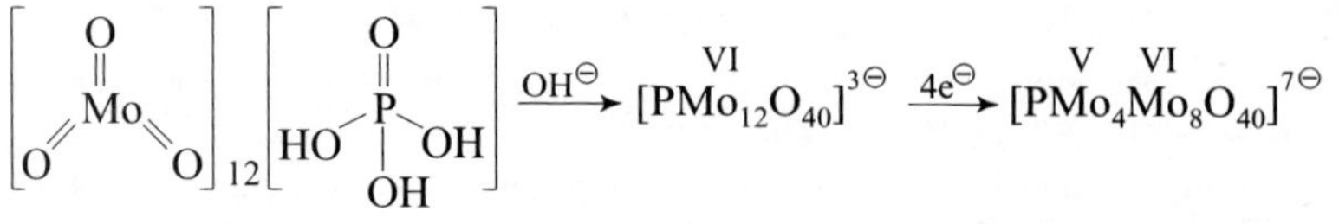

$$\left[\overset{O}{\underset{O}{\overset{\|}{Mo}}}\right]_{12} \left[HO-\overset{O}{\underset{OH}{\overset{\|}{P}}}-OH\right] \xrightarrow{OH^{\ominus}} [PMo_{12}O_{40}]^{3\ominus} \xrightarrow{4e^{\ominus}} [PMo_4Mo_8O_{40}]^{7\ominus}$$

Phosphomolybdic acid

Phosphomolybdenum blue
(absorbs at 660 nm)

The assay is actually performed in two steps. First, protein is reacted for 10 minutes, at room temperature, with alkaline copper sulphate (AR). During this incubation, a tetradentate copper complex forms from four peptide bond nitrogen atoms and one atom of copper (this is the "Biuret reaction"). Second, a phosphomolybdic-phosphotungstic acid solution is added. This reagent gets reduced due to the production of molybdenum blue and tungstun blue, producing an intense blue color, on incubation at room temperature. It is believed that the color enhancement occurs due to chelation of copper in a tetradentate complex. This helps in transfer of electrons from aromatic amino acids to the phosphomolybdic-phosphotungstic acid complex. The blue color continues to intensify during the 30 minutes incubation at room temperature. It has been suggested that during this period, a rearrangement of the initial unstable blue complex leads to the stable blue colored complex which has higher absorbance (Lowry, *et al.*, 1951; Legler, *et al.*, 1985). This blue color is optimally measured at 750 nm (peak at 750 nm and a shoulder at 660 nm), but it can be measured at any wavelength between 650 nm and 750 nm, with little loss of color intensity.

The presence of any of the five amino acid residues like tyrosine, tryptophan, cysteine, histidine and asparagine in the peptide or protein backbone enhances the amount of color produced, because they contribute additional reducing equivalents to further reduce the phosphomolybdic/phosphotungstic acid complex. For small peptides, the amount of color increases with the size of the peptide.

Note

- With the exception of tyrosine and tryptophan, free amino acids do not produce a colored product with the Lowry reagent.
- The reaction with Folin- Ciocalteu reagent depends on pH and a working range of pH 9-10.5 is essential for developing the intense blue colour.

Advantages of using Folin's Reagent: The measurement of protein with Folin's reagent has the following advantages:

1. It is simpler as well as much easier to adapt for small scale analyses.
2. It is a highly sensitive assay.
3. It is 10 or 20 times more sensitive than measurement of the ultraviolet absorption at 280 nm or ninhydrin reagent or Biuret reaction.
4. It is more specific and less liable to disturbance by turbidities.

Disadvantages of Folin's Reagent: Following are the disadvantages of Folin reagent(FR):

1. FR is highly sensitive to light.
2. The amount of color is different for different proteins based on the composition of characterising amino acids. The color is not always proportional to concentration.

In spite of all these concerns, Folin's method can be used in the measurement of protein during enzyme fractionations, mixed tissue proteins, measurement of very small absolute amounts of protein or highly diluted protein and analyses of large number of similar protein samples.

Why Bovine Serum Albumin (BSA) is used as standard: BSA is commonly used to determine the quantity of other proteins by comparing an unknown quantity of protein to known the amounts of BSA. Constructing a protein standard curve is of prime importance for studying the activity of an enzyme/protein, as analysis relies on accurate quantitation of protein concentration. BSA is used because of the following advantages:

1. Its ability to increase signal in assays.
2. Lack of effect in many biochemical reactions.
3. Easy availability in pure form.
4. Low cost, as it can be easily obtained from cattle blood.
5. This protein does not affect other enzymes which do not need it for stabilization.

Experimental Procedure

- Take a set of 8 clean and dry tubes and mark them as blank, T_1 to T_6 and TEST respectively. Label the pipettes for BSA, AR, FR and water.
- Add water and then suitable aliquots of BSA solution (0, 20, 40, 60, 80, 100 µg) respectively in blank to T_6 test tubes, as per the Table 3.2 below, such that the total volume becomes 1.0 mL (Do not add BSA to blank test tube).
- Add 1.0 mL of the given test solution, undiluted, to the test tube labelled as TEST.
- Now, add 2.5 mL of analytical reagent to each of the 8 test tubes. Shake the contents immediately and leave the test tubes undisturbed for 10 minutes. The solution turns light blue due to formation of tetradentate copper (II) complex.
- Next, add 0.25 mL of Folin's reagent to all the 8 test tubes. Mix the contents properly and allow the color to develop in dark for 30 minutes. (cover the beaker containing test tubes with filter paper and keep away from sunlight). After 30 minutes, set the zero (base line) of the instrument at 660 nm, with test tube labelled blank (Test tube 1).
- Next, record the absorbance of the known and test sample.
- Plot a standard curve of A v/s µg/mL of BSA at 660 nm. Use this curve to determine the amount of protein in the test sample of BSA/ protein.

Observation and Calculation

Table 3.2 BSA Standard Curve

S. No.	Water (mL)	BSA (mL)	Concentration of BSA (µg/mL)	AR Reagent (mL)	INCUBATE 10 MIN	Folin's Reagent (in mL)	INCUBATE 30 MIN*	Absorbance (OD) at 660 nm
1.	1.0	0.0 (Blank)	0	2.5		0.25		NIL (Use this solution to set instrument to zero)
2.	0.9	0.1	20	2.5		0.25		
3.	0.8	0.2	40	2.5		0.25		
4.	0.6	0.4	80	2.5		0.25		
5.	0.4	0.6	120	2.5		0.25		
6.	0.2	0.8	160	2.5		0.25		
7.	0.0	1.0	200	2.5		0.25		
8.	X	Test 1		2.5		0.25		

*In dark

Note

- Although BSA is a water-soluble protein, it takes time to dissolve completely in water. So, the stock solution should be prepared and kept for at least an hour before starting the experiment (The reagent is stable for 1 month).

- While dissolving the protein, do not shake the protein solution vigorously. (This prevents frothing, which can lead to denaturation of protein)

- The Lowry solution (or analytical reagent) should be prepared fresh, on the day of estimation. Though the individual solutions required to prepare the Lowry solution can be prepared in advance and mixed on the day of the estimation.

- Folin's reagent (1:1 mixture with water) should also be prepared fresh. It is light sensitive, so it should be prepared at the time of addition in the experiment and should be kept in an amber or dark colored container or flask covered with silver foil.

- Contents of the tube are to be mixed immediately after addition of Folin's reagent. Proper color development is affected if this step is unduly delayed.

- Compounds like Tris, EDTA, phenols, sugars and inorganic ions like K^+, NH_4^+ and Mg^{2+} interfere in this procedure. Detergents and chelating agents precipitate copper and prevent its complexation with peptide.

- Pipetting should be done very carefully to avoid errors.
- Add water to the test tubes before adding BSA, as glass can coagulate the protein.
- Reagent should always be in excess so that all the protein combines with the reagent.
- The amount of color produced in this assay by any given protein depends on the amino acid composition of the protein. Therefore, two different proteins, for example, at concentrations of 200 µg/mL each, can give different color yields in this assay. It must be appreciated, therefore, that using BSA or any other protein for that matter as standard gives only an approximate measure of the protein concentration.
- Test solutions prepared may be 150 µg/mL or 100 µg/mL.

Utility of This Exercise

- Protein quantization is necessary before processing protein samples for separation, isolation and analysis by chromatography, electrophoresis and immunochemical techniques.
- Proteins are also useful for therapeutic, medicinal and diagnostic purposes, which again require the protein in use to be estimated.
- All enzymes are proteins. However, the whole of protein present in a given sample need not to be active as enzyme. Thus, by measuring the specific activity (S) of the enzyme we can predict the amount of protein in the sample which actually shows catalytic activity.

$$S = U/\text{mg of protein/mL}$$

where, U is the activity of enzyme obtained from activity studies and *mg of protein/mL* is obtained from Lowry's estimation.

Lessons to Take Home

After doing this exercise, the students will be able to do the following:

- Estimate the amount of protein present in an unknown sample.
- Understand how absorbance of a sample varies with concentration.
- Understand the role of alkaline pH in Lowry estimation
- Understand how Lowry estimation is superior to Biuret reaction.

3.4 ENZYME ACTIVITY

Living organisms are able to obtain and use energy rapidly because of the presence of biological catalysts called **enzymes**. Enzymes are mostly proteins except ribozymes, which are nucleic acids. Their catalytic power is high, often greater than that of synthetic catalyst. As with the inorganic catalysts, enzymes change the rate of a chemical reaction but do not affect the final equilibrium. Also, they are only required in small amounts to bring about the transformation for a large number of molecules. However, unlike most inorganic catalyst, enzymes operate in aqueous conditions and at physiological pH and temperature. They have a very narrow specificity range *i.e.* they will only catalyze a comparatively small range of reactions or in some cases, only one reaction.

Why study enzymes?

Enzymes have numerous practical applications. Deficiency of an enzyme or its absence may lead to genetic disorders. The increased activity of the enzyme on the other hand can also lead to abnormal conditions. Besides their usefulness in medicines, enzymes are becoming important tools in food processing, laundry, agriculture etc.

What type of reactions do the enzymes catalyze: According to International Commission for Enzymes, enzymes catalyse six types of reactions: Redox reactions, rearrangement of bonds (isomerization), tranfer of functional groups, hydrolysis, bond making and bond breaking.

How do enzyme catalyze reactions: Normal reactions in the lab are catalysed by

1. Changing pH
2. Increasing temperature
3. Addition of catalyst-acids, alkali, etc.

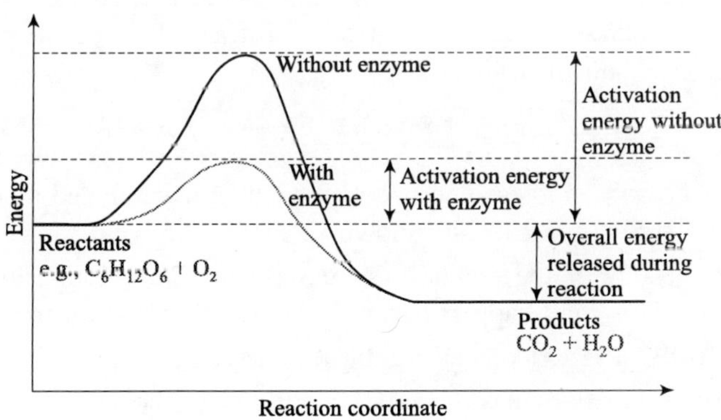

Fig. 3.4 Enzyme Catalysed Reaction

Biomolecules are sensitive to changes in temperature and pH. Therefore, their reactions cannot be catalysed by inorganic catalysts. Enzymes catalyze the biomolecular reactions, at specific temperature and pH, by reducing the activation energy of the reaction (Figure 3.4). They do so by providing an alternate pathway for the reaction.

Some enzymes are simple proteins which act on their own e.g. trypsin, pepsin etc. whereas other enzymes like salivary amylase require a non-protein part (called **cofactor**) to exhibit catalytic activity (Figure 3.5). A cofactor can be **inorganic ion,** e.g., Fe^{2+}, Mg^{2+}, Ca^{2+}, Zn^{2+} or an organic molecule. The organic molecules like vitamins which are loosely bound to the enzyme are called **co-enzymes** whereas the organic molecules like nucleic acids and lipids which are tightly bound to the enzyme are called **prosthetic groups**. The cofactors help enzymes by providing catalytic groups required for binding and catalysis.

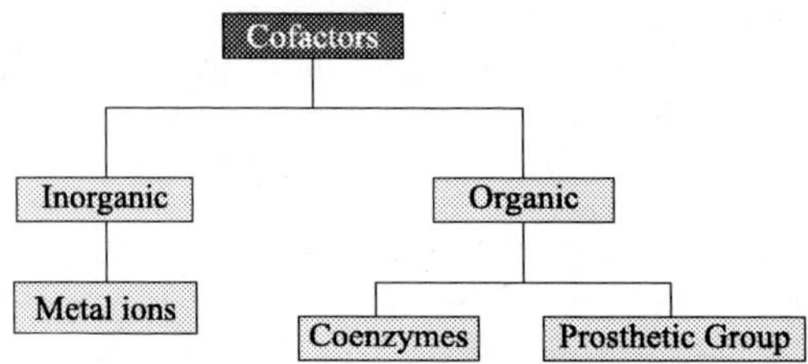

Fig. 3.5 Types of Cofactors

The active enzyme-cofactor complex is called **holoenzyme**. Whereas, the inactive protein part left after the removal of cofactor is called **apoenzyme**. The enzyme-cofactor complex/simply enzyme binds the substrate at the **active site** and helps in the formation of the product through binding and catalysis. The final product has no affinity for the enzyme surface. Hence, it leaves the enzyme surface making way for the fresh substrate. This entire process may be represented as shown below:

$$E + S \rightleftharpoons ES \rightleftharpoons EP \longrightarrow E + P$$

Measurement of reaction rates: The rate of any enzyme catalyzed reaction shows a "progress curve" similar to one shown in Figure 3.6.

Clearly, the rate of the reaction decreases with time though it is linear at the beginning. This may happen because of the following reasons:

(a) Products of reaction may inhibit the enzyme.

(b) Reverse reaction may become important as the concentration of the product increases.

(c) Degree of saturation of enzymes with substrate or coenzymes may fall because of their conversion to the product.

(d) Enzyme or coenzyme may undergo some denaturation at the temperature or pH of the reaction, due to instability.

The initial reaction rate is measured when none of these factors have become operational.

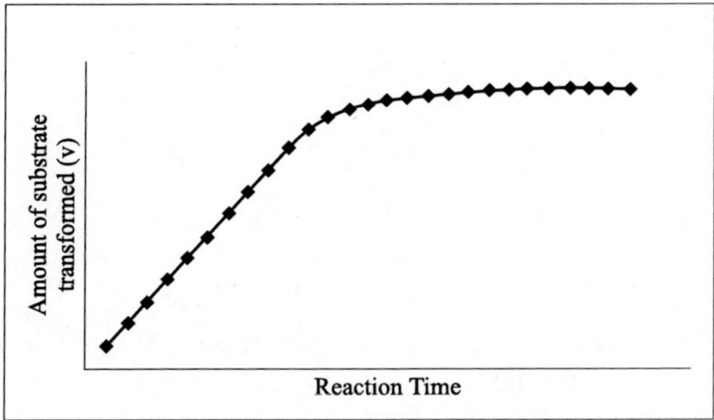

Fig. 3.6 Progress Curve

The enzyme activity depends on a large number of factors such as substrate/ enzyme concentration, pH, temperature, and presence of activator/inhibitor. It is therefore important that all reagents attain the required temperature and pH before commencing the enzymatic reaction. The enzymatic reaction should be carried out in thermostat condition and adequate buffer should be added to ensure that changes in pH and temperature are not too high (as most of the enzymes are active between 40–50°C; pH 5-9). Ionic strength, pH and temperature chosen for the reaction should be in the range where enzyme is most stable. A detailed discussion on factors affecting an enzyme catalyzed reaction is taken up in *section 3.4.2*.

From the above discussion it can be concluded that the optimal conditions required to study an enzyme catalyzed reaction are as follows:

1. *Optimum pH:* pH of the medium should correspond to optimal pH of the enzyme.
2. *Unlimited substrate*: Substrate should be unlimited so that all active sites of enzymes are saturated.
3. *Activator/cofactor/coenzyme:* Should be present to provide catalytic groups.
4. *Optimum enzyme concentration*
5. *Optimum temperature*: Incubation should be done at the optimum temperature for enzyme.
6. *Incubation time*: Should be where linear kinetics is followed (Fig. 3.6).
7. *Termination of reaction*: Reaction should be *terminated* with heat/acid/base.

Enzyme Activity: Enzyme activity is measured in terms of enzyme units (U). 1 enzyme unit is the amount of enzyme that catalyzes conversion of 1μ mole of substrate into product per minute, under specified conditions of temperature and pressure.

$$1\,U = 1\,\mu\ mol\ min^{-1}$$

Techniques to Measure Enzyme Activity

Rate of utilization of substrate or rate of formation of the product is measured for quantitative estimation of enzyme activity. Activity of an enzyme can be determined in three different ways in the lab. These are:

(a) *Sampling method:* Samples are withdrawn at regular intervals from the reaction mixture and analyzed for disappearance of the substrate or appearance of the product.

(b) *Continuous method:* Changes in concentration of substrate, coenzyme or product are measured continuously while the reaction is in progress. This will require hyphenated techniques.

(c) *Discontinuous method:* The reaction is allowed to proceed for a specified period of time and then it is stopped using suitable reagents. The amount of product liberated or reactant consumed can then be measured. All enzyme assays discussed in this book are based on this method only.

In all the methods, a controlled assay (which is generally same as test assay in all respects except that it lacks either E or S) should be simultaneously carried out. Any changes in the control assay would give an idea of the background reaction. Some of the commonly used analytical methods for determining either the amounts of substrate or product or coenzyme in the assay are given below:

1. UV and visible spectrophotometric method
2. Luminescence method
3. Radioisotopic method
4. Microcalorimetric method
5. Florimetrically

Note: Spectrophotometric method, using UV-Visible range of electromagnetic spectrum, is most popular at the UG level for enzyme assays, amongst all the methods listed above.

3.4.1 Study the Action of α-Amylase on Starch at Room Temperature

Basic Lab Operations: Spectrophotometry.

Estimated Time: 120 minutes

Chemicals Required: Maltose solution (1 mg/mL in water); 2,4-dinitrosalicylic acid (DNSA reagent, Appendix II), acetate buffer (pH 5.6, Appendix-II), α-amylase solution [1 mg/mL in acetate buffer (pH 5.6), diluted 1:50 with acetate buffer to get working solution of 20 μg/mL]; 1% starch solution in acetate buffer, 1% NaCl solution in acetate buffer.

What You Should Know: All living organisms require energy to live and do the work. We get this energy from the food we consume. Food is digested in the body through various metabolic pathways to generate glucose which undergoes glycolysis, Kreb cycle and oxidative phosphorylation to generate energy in the form of ATP molecules and also liberate CO_2 and water as by-products. Digestion of complex carbohydrates like starch starts in our mouth. Our teeth cut the food into smaller fragments and saliva secreted by salivary glands mixes with these smaller fragments of food. Enzyme, salivary amylase, present in saliva breaks down the starch to form maltose (Figure 3.7). Maltose is then broken down to glucose *via* action of maltase.

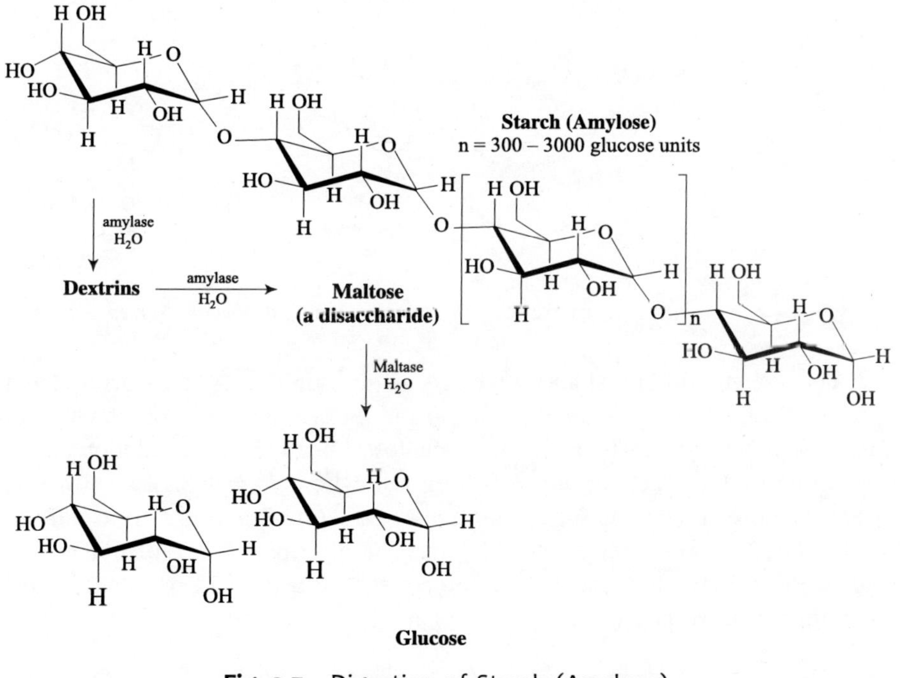

Starch (Amylose)
n = 300 – 3000 glucose units
amylase
H_2O
Dextrins
amylase
H_2O
Maltose
(a disaccharide)
Maltase
H_2O
Glucose

Fig. 3.7 Digestion of Starch (Amylose)

Amylases are hydrolases, acting on α-1,4-glycosidic bonds. They can be further subdivided into α-, β- and γ-amylases. α-Amylase (AAM) is an endoenzyme that acts as catalyst for the hydrolysis of α-linked polysaccharides such as starch and glycogen, yielding glucose and maltose. α-Amylase is found in the human body in pancreatic juice and salvia. It requires chloride/bromide ion for its catalytic activity (**cofactor**). Calcium ions are required to hold the three-dimensional structure of the protein (**metalloenzyme**). In the absence of calcium ions, the protein unfolds and loses its catalytic activity. Salivary amylase is inactivated in the stomach by gastric acid (pH 3.3). A general mechanism of action of alpha amylases is shown in Figure 3.8, taking salivary amylase as example.

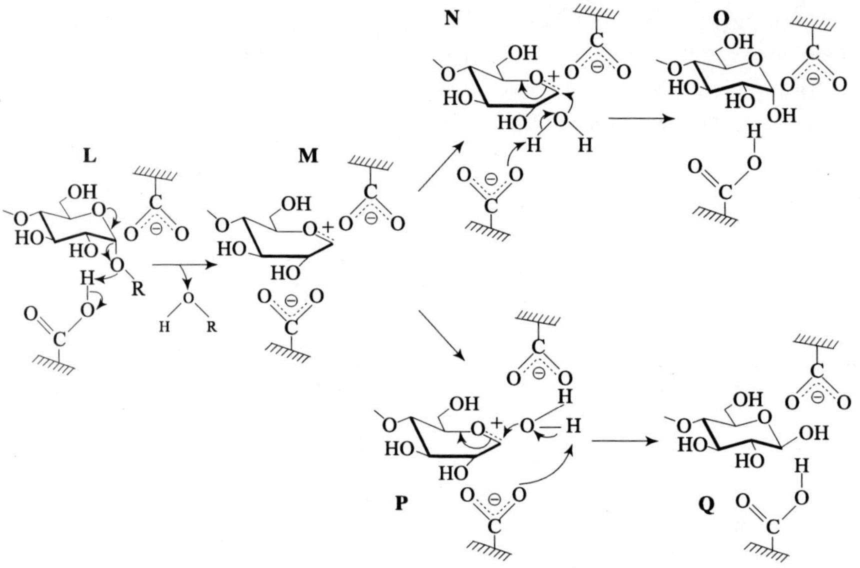

Fig. 3.8   Mechanism of Action of Salivary Amylase in the Active Site

The activity of an alpha amylase can be studied by measuring the μ moles/mL/min of maltose liberated from the known quantities of starch, upon action of the enzyme. The amounts of maltose liberated in the solution can be quantified by the reaction with 3,5-dinitrosalicylic acid (DNSA), an oxidising agent. Maltose, a reducing sugar, will reduce DNSA to 3-amino-5-nitro salicylic acid, which absorbs strongly at 540 nm. The absorbance at 540 nm will be proportional to the amount of 3-amino-5-nitro salicylic acid produced, which is in turn proportional to the amount of maltose liberated.

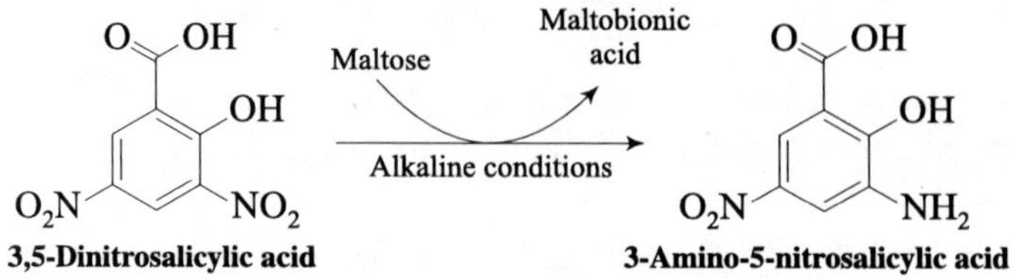

Role of Ca²⁺ ions: Salivary amylase is a metalloenzyme which requires Calcium(II) ion for its activity. Calcium ion stabilizes the $3°$ structure of enzyme and in the process exposes active site to substrate.

Role of Rochelle salt in DNSA:

1. Reduces tendency of solution to dissolve oxygen by increasing ionic concentration. Oxygen is capable of oxidizing reducing sugar and introducing errors.

Role of DNSA:

1. Arrests the reaction towards the end by changing pH of the reaction.

2. Helps quantification of enzyme activity by oxidation of maltose.

Role of Chloride ions

1. It is an allosteric effector. It binds to active site of the enzyme and accelerates the reaction by:

   (a) Increasing pK_a for ionisation of COOH group of Glu-233, present in the active site, such that it is unionised at pH of maximum enzyme activity. It does so by neutralising the partial positive charge on the carbonyl carbon, such that carboxyl group OH is not able to enter into resonance with carbonyl group.

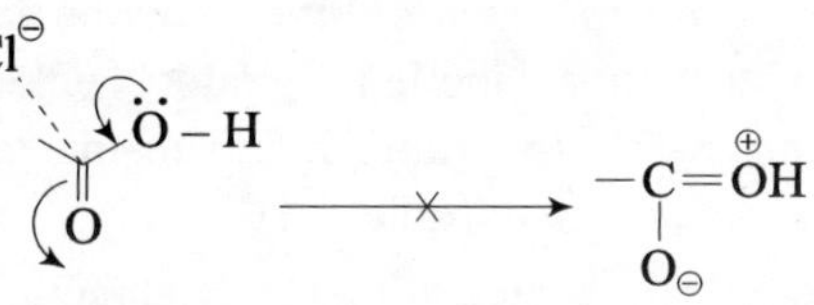

   (b) Carboxyl group of Glu-233 is required for H⁺ donation in amylolytic cleavage of 1,4-glycosidic linkage.

Experimental Procedure

Step 1: Plot maltose standard curve as per details given below:

- Take a set of 9 labelled test tubes. Label the first one as blank and remaining to be numbered 1 to 8.
- To each test tube, add maltose except in the blank (See Table 3.3).
- Add distilled water to each tube to make up the volume to 1.0 mL.
- Next add DNSA (1.0 mL) to each test tube and shake. Cover the test tubes with aluminium foil and put the test tubes in boiling water bath for 10 minutes.
- Cool and dilute with distilled water (4.0 mL).
- Set the spectrophotometer at 540 nm. Set the spectrophotometer to zero using blank.

- Record the absorbance for test samples sequentially at 540 nm.
- Plot a standard curve of absorbance at 540 nm v/s concentration of maltose in µg/mL (See figure 3.9).

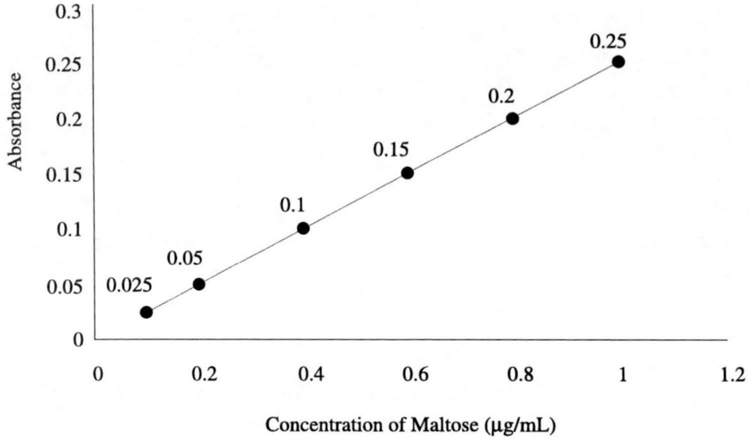

Fig. 3.9 Maltose Standard Curve

Step 2: Carry out the assay as per detail given below:

- *Prepare a control sample*: Pipette the buffer solution (1.4 mL) in a clean and dry test tube labelled C, add NaCl (0.1 mL) followed by starch (0.5 mL) to this test tube. Incubate the test tube at room temperature for 20 minutes (See Table 3.4).

- *Prepare a test sample*: Take buffer solution (1.2 mL) in a clean and dry test tube, labelled T, and add NaCl (0.1 mL) to it and shake. Then add enzyme solution (0.2 mL) and shake. Incubate the mixture at room temperature for 5 min. Next add starch (0.5 mL), shake and incubate the mixture again at room temperature for 20 minutes.

- Now add DNSA (1.0 mL) to both test and control samples and shake.

- Take the test tubes containing test and control samples, cover with foil and keep in boiling water bath for 10 minutes. Cool the mixtures and add distilled water (3.0 mL) to both the samples and shake.

- Set the spectrophotometer to zero with the control and then record the absorbance of the test sample at 540 nm.

- Put the absorbance value in the maltose standard curve and determine the concentration of maltose in µg/mL.

- Calculate the activity of alpha amylase in µmol/mL/min.

Observation and Calculations

Table 3.3   Maltose Standard Curve

S. No.	Maltose (mL)	Concentration of Maltose (μg/mL)	Distilled Water (mL)	DNSA (mL)			Absorbance at 540 nm
1.	0.0 (Blank)	-	1.0	1.0			Nil (Use this solution to set instrument to zero).
2.	0.2	200	0.8	1.0	Keep in boiling water bath for 10 min.	Cool, dilute with 4.0 mL distilled water	
3.	0.3	300	0.7	1.0			
4.	0.4	400	0.6	1.0			
5.	0.5	500	0.5	1.0			
6.	0.6	600	0.4	1.0			
7.	0.7	700	0.3	1.0			
8.	0.8	800	0.2	1.0			
9.	0.9	900	0.1	1.0			

Note: Molar mass of maltose = 342 g/mol

Table 3.4   Control Sample

S. No.	Buffer (mL)	1% NaCl (mL)	Enzyme (mL)	1% Starch (mL)		DNSA (mL)			Absorbance at 540 nm
Control I	1.4	0.1	0.0	0.5	Incubate R.T. for 20 min.	1.0	Cover and keep in boiling water bath for 10 min.	Cool, dilute with 3 mL distilled water	Nil (Use this solution to set instrument to zero).

Table 3.5   Test Sample

S. No.	Buffer (mL)	1%NaCl (mL)	Enzyme (mL)		1% Starch (mL)		DNSA (mL)	Cover and keep in boiling water bath for 10 min.	Cool, dilute with 3 mL distilled water	Absorbance at 540 nm
Test I	1.2	0.1	0.2	Incubate at R.T. for 5 min.	0.5	Incubate R.T. for 20 minutes	1.0			

Calculations:

Enzyme activity =

$$\frac{\text{Concentration}\left(\frac{\mu g}{mL}\right) \times \text{Dilution factor} \times \text{Total assay volume}}{\text{Enzyme (volume)} \times \text{Incubation time} \times \text{Molar mass of maltose}}$$

Volume of enzyme added = 0.2 mL

Dilution Factor = 50 (stock solution is diluted 1:50)

Total assay volume = 2.0 mL

Incubation time = 20 minutes

$$\text{Enzyme activity} = \frac{C\left(\dfrac{\mu g}{mL}\right) \times 50 \times 2.0}{0.2 \times 20 \times 342}$$

Therefore,

Enzyme activity $= C \times 0.0731\ \mu\ mol/mL/min$

Note

- Record the room temperature as activity of enzyme varies with temperature.
- Total assay volume of test, control and that in maltose standard curve should be the same.
- Control is enzyme blank in this experiment.
- Alpha amylase, depending upon different sources, may exhibit different optimum pH (normally 6.5-7.0). Ensure the optimum pH of the sample before measuring enzyme activity.
- α-Amylase solution: See Appendix II
- Acetate buffer (pH 5.6): See Appendix II
- Maltose solution: See Appendix II.
- 1% NaCl solution: See Appendix II

Result: The activity of alpha amylase at -------- °C was found to -------- μmol/mL/min.

Utility of This Exercise

Action of α-amylase on starch forms the basis of many industrial and commercial applications of amylase. Some of these are:

- α- and β-Amylases are used in brewing beer and liquor, which uses starch derived sugars as substrate.
- Amylases are used in bread making and to break down complex sugars such as starch (found in flour) into simple sugars.
- Bacilliary amylase is used in clothing and dishwasher detergents to dissolve starches from fabrics and dishes.
- Blood serum amylase may be measured for purposes of medical diagnosis. A higher than normal concentration may reflect one of the several medical conditions, including acute inflammation of pancreas.

Lessons to Take Home

After doing this exercise, the students will be able to do the following:

- Understand how carbohydrate digestion occurs in the human body.
- Understand why humans can digest starch and not cellulose.
- Understand the role of calcium and chloride ions in amylase activity.
- Explain why the enzyme activity is arrested on adding DNSA.
- Extend this experiment to determine sugar contents of fruits and fruit juices.

3.4.2   Study the Effect of Temperature on Activity of α-Amylase

Basic Lab Operations: Spectrophotometry.

Estimated Time: 120 minutes

Chemicals Required: Maltose solution (1.0 mg/mL in water); 2,4-dinitrosalicylic acid (DNSA reagent, See Appendix II); acetate buffer (pH 5.6, See Appendix II), α-amylase solution (See Appendix II); 1% Starch solution in acetate buffer, 1% NaCl solution in acetate buffer.

What You Should Know: Activity of an enzyme largely depends on the following five factors: Enzyme concentration, substrate concentration, temperature, pH, and presence of inhibitor/activator. Effect of these factors on enzyme activity is discussed in brief in this section:

1.   *Enzyme Concentration:* The rate of an enzyme catalyzed reaction is directly proportional to enzyme concentration (Fig. 3.10), *i.e.,* two molecules of enzyme act independently in solution and will act on twice as much substrate in a given time as compared to one molecule.

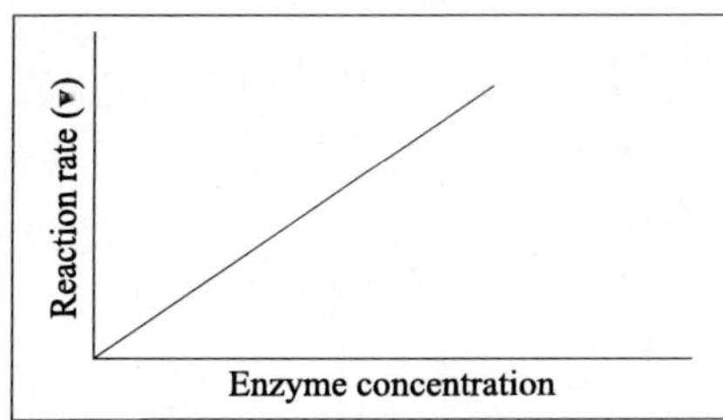

Fig. 3.10   Effect of Enzyme Concentration on Reaction Rate

There can be however, deviation from linearity due to some reasons. It is therefore important to establish the relationship between enzyme concentration and reaction rate before studying the enzyme kinetics. The factors affecting the linearity are:

(a) Presence of toxic impurity in any of the components of the mixture being incubated.

(b) Presence of dissociable activator or coenzyme in the enzyme preparation. (percentage of enzyme in active form will increase in such a case as the activator concentration increases).

(c) Limitation in measuring capacity of method being used rather than decrease in activity of enzyme.

(d) Greater affinity of enzyme for one of the components of the mixture.

(e) Presence of reversible inhibitor in the enzyme preparation (% enzyme in active form decreases due to formation of enzyme-inhibitor complex).

In spite of these factors, normally linear curves are obtained for rate of reaction versus concentration of E, all other factors remaining constant, and this forms the basis of all enzyme kinetic studies.

2. *Substrate Concentration:*   One of the most important factors affecting the rate of an enzyme catalyzed reaction, *in vitro*, is the concentration of the substrate. Variation of the reaction rate with concentration of the substrate [S] is shown by Michaelis-Menten curve (Figure 3.11). It is observed that at relatively low concentration of the substrate, reaction rate increases linearly with substrate concentration. This is due to fact that the enzyme is not saturated with the substrate. On increasing the [S], more and more substrate binds to the active site, hence rate of the reaction increases. (**Note:** Reaction in this part of the curve follows the first order kinetics). At higher substrate concentration, reaction velocity increases slowly (showing mixed order kinetics) finally, a point is reached when the velocity has attained a maxima (V_{max}), increasing substrate concentration beyond that does not alter the rate of the reaction, because the enzyme has become saturated (zero order kinetics), so only substrate required now is to replace the product, which leaves the active site after formation.

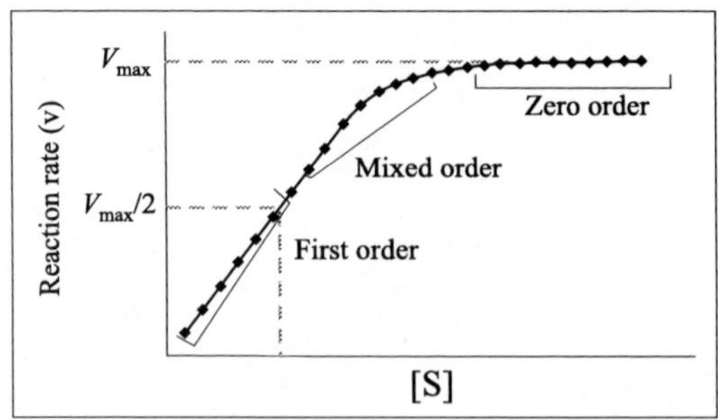

Fig. 3.11   Variation of Reaction Rate with Substrate Concentration

3. *Effect of pH:* Enzymes are active over a narrow range of pH (usually 5-9) and a plot of enzyme activity against pH is usually a bell-shaped curve as shown in Fig. 3.12.

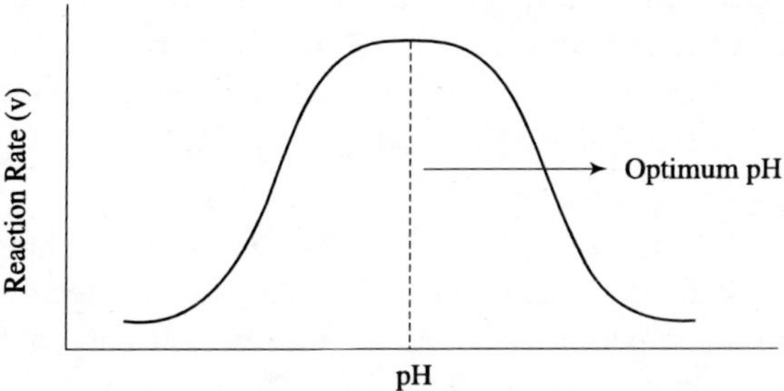

Fig. 3.12   Effect of pH on the Rate of Enzyme Catalysed Reaction

The pH at which an enzyme shows maximum activity is called the **"optimum pH"** of the enzyme. This pH is characteristic of an enzyme and depends on the structure of the enzyme. Variation of enzyme activity with pH is due to change in the state of ionization of the protein and also other components of the mixture. Michaelis and Davidsonn, in 1911, suggested that protein is active only in one of its many ionized forms, so that a change in pH on either side of the optima produces a decrease of this form and hence the activity falls. Therefore, decrease in the reaction rate on the alkaline side of the curve may be due to denaturation/destruction of the enzyme and on acidic side due to decreased concentration of the active form of enzyme.

4. *Temperature:* For any molecule to react, it is important that it has a certain amount of energy of activation. Enzymes, we know, function by lowering this energy of activation thereby facilitating the reaction to proceed smoothly at low temperatures. The overall energy of the reaction is however unaffected by the enzyme. The rate of an enzyme catalyzed reaction increases with temperature till it reaches the maximum. The velocity decreases on further increasing the temperature as shown in Figure 3.13.

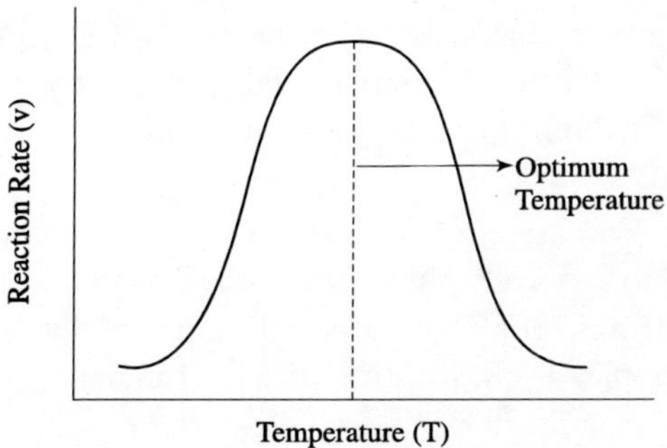

Fig. 3.13   Effect of Temperature on the Rate of Enzyme Catalysed Reaction

Initial increase in the reaction rate with increase in temperature can be explained due to increase in the fraction of molecules having energy higher than activation energy. The temperature at which enzyme activity is maximum, is called "**optimum temperature**" of the enzyme. Activity of enzyme falls beyond this temperature due to denaturation. Decrease in activity could also be because change in temperature can affect (a) stability of enzyme, (b) ES affinity, (c) pK_a values of the components of the reaction may get altered due to the effect on heat of ionization, and (d) affinity of enzyme for activator/ inhibitors.

Most enzymes are active in the narrow temperature range of 40-50°C. Effect of temperature on enzyme stability can be studied by exposing the enzyme to various temperatures for a definite period and then measuring its activity at optimum temperature. The effect of affinity could be eliminated by using sufficiently high concentration of substrate/activator to saturate the enzyme. This section discusses the effect of temperature on activity of alpha amylase.

Experimental Procedure

- Prepare acetate buffer (pH 5.6): See Appendix II.
- Draw a maltose standard curve (Refer *section 3.4.1*)
- *Prepare test solutions for each temperature:* Take six clean and dry test tubes. Label test tubes, T-1 to T-6. Pipette buffer solution (1.2 mL, see Table 3.7), 1% NaCl (0.1 mL) and the enzyme solution (0.2 mL) in the test tubes.

- *Prepare control solutions for each corresponding temperature*: Take six clean and dry test tubes. Label the test tubes, C-1 to C-6. Now pipette out buffer solution (1.4 mL, see Table 3.6), 1% NaCl (0.1 mL) into each of these test tubes.

- *Pair six test samples with corresponding controls:* (For example: pair C-1 and T-1, C-2 and T-2 and so on). Incubate each pair of control and test at a pre-identified temperature (temperatures can be 10, 20, 37, 50, 60 and 70°C) for 5 minutes. Add starch solution (0.5 mL) to each of the test and control solutions. Incubate the test tubes further at respective temperatures for 20 minutes. Add DNSA (1.0 mL) to each of the test and control samples and shake. Cover the test tubes and keep them in boiling water bath for 5 minutes, then cool and dilute with 3.0 mL water.

- *Set the spectrophotometer to zero*: Use the control (C-1) to set the instrument to zero and then record the absorbance of the test sample (T-1) at 540 nm. Similarly, proceed to record the absorbance of other five test samples by first setting the instrument to zero with respective controls.

- Put the absorbance values in the maltose standard curve and determine the concentration of maltose, in μg/mL, at six different temperatures. Calculate the activity of salivary amylase in μmol/mL/min. at each of these temperatures. Then plot the bell-shaped graph of rate of reaction v/s temperature. Use the curve to determine the temperature at which activity of the enzyme is maximum. This temperature is the optimum temperature of the enzyme.

Observations and Calculations

Table 3.6   Control

S. No.	Temperature (°C)	Acetate Buffer (mL)	1% NaCl (mL)	Enzyme (mL)		1% Starch (mL)		DNSA (mL)	Cover, keep in boiling water bath for 5 min. Cool and dilute with 3.0 mL water	Absorbance (540 nm)
1.	10				Incubate at respective temperature for 5 min.		Incubate at respective temperature for 20 min.			NIL*
2.	20									NIL
3.	37	1.4	0.1	0.0		0.5		1.0		NIL
4.	50									NIL
5.	60									NIL
6.	70									NIL

*Use these to set instrument to zero at respective temperatures

Table 3.7   Test Sample

S. No.	Temperature (°C)	Acetate Buffer (mL)	1% NaCl (mL)	Enzyme (mL)		1% Starch (mL)		DNSA (mL)	Keep in boiling water bath for 5 min. Cool and dilute with 3.0 mL water	Absorbance (540 nm)
1.	10				Incubate at respective temperature for 5 min.		Incubate at respective temperature for 20 minutes			
2.	20									
3.	37	1.2	0.1	0.2		0.5		1.0		
4.	50									
5.	60									
6.	70									

Enzyme activity at T°C =

$$\frac{\text{Total assay volume} \times C \times \text{Dilution factor}}{\text{Volume of Enzyme} \times \text{Incubation time} \times \text{Molar mass of maltose}} \; \mu\text{mol/mL/min}$$

$$= C \times 0.0731 \; \mu\text{mol/mL/min}$$

where, C is concentration of maltose in μg/mL at respective temperature. Molar mass of maltose is 342 g/mol. For details, see *section 3.4.1*.

Result: The effect of temperature on the activity of α-amylase was studied. The optimum temperature for the enzyme was found to be -------------- °C.

Utility of This Exercise See *section 3.4.1*

Lessons to Take Home

After doing this experiment, the students will be able to do the following:

- Understand the factors which affect the rate of enzyme catalysed reaction.
- Understand why reaction rate rises initially with rise in temperature and falls beyond optimum temperature.
- Determine optimum temperature for other enzyme formulations.
- Understand how some organisms survive in cold waters of Alaska, some in plains and some in hot springs.

3.4.3   Study the Effect of pH on Activity of α-amylase at Room Temperature

Basic Lab Operations: Spectrophotometry.

Estimated Time: 120 minutes

Chemicals Required: Maltose solution (1 mg/mL in water); 2,4-dinitrosalicylic acid (DNSA reagent, see Appendix II); acetate buffer (pH 4.5 and 5.5), phosphate buffer (6.0, 6.5,7.0, 7.5 and 8.0), α-amylase solution: (prepared in respective buffers, see Appendix II); 1% starch solution in respective buffers, 1% NaCl solution in respective buffers.

What You Should Know: For detailed on theory, please *see section 3.4.1 and 3.4.2.*

Experimental Procedure

- Prepare the required buffer solutions.
- Plot the maltose standard curve: (*see section 3.4.1*)
- *Prepare test solutions for each pH:* Take six clean and dry test tubes. Label test tubes, T-1 to T-6. Pipette buffer solution of suitable pH (1.3 mL, see Table 3.8), 1% NaCl (0.1 mL) and the enzyme solution (0.1 mL) in the test tubes. Incubate at room temperature/ 37°C for 5 minutes. Then add 1% starch (0.5 mL) and again incubate at room temperature/ 37°C for 20 minutes.
- *Prepare control solutions for each corresponding pH:* Take six clean and dry test tubes. Label the test tubes, C-1 to C-6. Now pipette out suitable buffer solution (1.4 mL, see Table 3.9), 1% NaCl (0.1 mL) and starch (0.5 mL) into each of these test tubes. Incubate for 20 minutes at room temperature/ 37°C.

- *Pair corresponding control and test samples*: Pair test and control samples corresponding to identical pH (for example, C-1 and T-1, C-2 and T-2 and so on----). Add 1.0 mL of DNSA to each of the test and control samples. Cover the test tubes with aluminium foil and keep in boiling water bath for 5 minutes. Cool the test tubes and dilute with 3.0 mL water. Set the spectrophotometer filter at 540 nm and set the instrument to zero using C-1. Take the absorbance reading at 540 nm using T-1.

- Repeat the same procedure for remaining five sets of controls and test.

- Use the absorbance values for each pH to determine the corresponding maltose concentrations from maltose standard curve.

- Determine the enzyme activity at each pH.

- Plot a graph of reaction rate (activity) v/s pH.

- Identify the optimum pH of the enzyme.

Observation and Calculations

Table 3.8   Test Sample

S. No.	pH	Buffer (mL)	1% NaCl (mL)	Enzyme (mL)		1% Starch (mL)		DNSA (mL)	Cover, keep in boiling water bath for 5 minutes. Cool and dilute with 3.0 mL water	Absorbance (540 nm)
1.	4.5	↑	↑	↑	Incubate at room temperature for 5 minutes	↑	Incubate at room temperature for 20 minutes.	↑		
2.	5.5									
3.	6.0	1.3	0.1	0.1		0.5		1.0		
4.	6.5									
5.	7.0									
6.	8.0	↓	↓	↓		↓		↓		

Table 3.9   Control

S.No.	pH	Buffer (mL)	1%NaCl (mL)	Enzyme (mL)		1% Starch (mL)		DNSA (mL)	Cover, keep in boiling water bath for 5 minutes. Cool and dilute with 3.0 mL water	Absorbance (540 nm)
1.	4.5	↑	↑	↑	Incubate at room temperature for 5 minutes	↑	Incubate at room temperature for 20 minutes	↑		NIL*
2.	5.5									NIL
3.	6.0	1.4	0.1	0.0		0.5		1.0		NIL
4.	6.5									NIL
5.	7.0									NIL
6.	8.0	↓	↓	↓		↓		↓		NIL

*Use these solutions to set instrument to zero.

Enzyme activity at respective pH =

$$C \times \text{dilution factor} \times \frac{\text{Total assay volume}}{\underset{\text{enzyme}}{\text{Volume of}} \times \underset{\text{time}}{\text{Incubation}} \times \underset{\text{of maltose}}{\text{Molar mass}}} \,\mu\text{mol/mL/min}$$

$$= C \times 0.1461 \;\mu\text{mol/mL/min (see } section \; 3.4.1 \text{ for details)}$$

where, C is concentration of maltose in μg/mL and molar mass of maltose is 342 g/mol.

Result: The activity of α-amylase was studied at ------------ °C and the optimum pH for the enzyme was found to be ------------ .

Utility of This Exercise: *See section 3.4.1*

Lessons to Take Home

After doing this exercise, the students will be able to do the following:

- Determine the optimum pH for the given enzyme
- Carry out enzyme assays at optimum pH of the enzyme.
- Explain why enzyme activity is maximum at optimum pH.
- Explain why enzyme is denatured above optimum pH
- Correlate the phenomenon like souring of milk to pH changes in protein samples.

Post Lab Questions

1. What is a pHmetric titration curve? What is the basic aim of such titrations?
2. In acid-base titrations, what is the need to draw a first derivative curve?
3. How many maxima will you get in the first derivative plot for an acidic/basic amino acid?
4. Why is it important to prepare the solution of glycine in a strongly acidic solution prior to titration.?
5. Why is it important to prepare glycine solution with matching normality as that of given HCl?
6. Can you dissolve glycine in water for determination of its pI value, pHmetrically?
7. Name two methods useful for protein estimation in the lab.
8. What is the need to estimate protein in the given sample?
9. Define activity and specific activity of an enzyme.
10. Name the best method for estimation of protein in the lab.
11. How is Lowry's method superior to Biuret method?
12. What is a buffer? Name two commonly used buffers in the lab.
13. Why BSA is chosen as a standard for protein estimation?

14. What is Folin's reagent? What precaution is to be taken while preparing the Folin's reagent?

15. What is the composition of AR in Lowry's estimation? Explain the role of Rochelle's salt in Lowry's method.

16. What is the advantage and disadvantage of Folin's reagent in protein chemistry?

17. What is a standard curve? Explain the need to draw a standard curve in biochemical experiments.

18. What are enzymes and why do we need to study them?

19. What do you mean by progress curve? Why do you need to draw them?

20. Why are biochemical assays carried out in thermostat conditions?

21. What are the different approaches commonly used to measure the enzyme activity? Which of this method did you use?

22. Which standard curve did you draw while measuring the activity of salivary amylase and why?

23. Name two types of amylases found in the body.

24. What is DNSA? Can it be used for estimation of any other type of biomolecules?

25. Explain the principle behind estimation of amylase activity.

26. Which buffer did you use while measuring amylase activity and why?

27. What is the role of NaCl in amylase activity measurement?

28. To which class of enzyme does amylase belong? What is the role of Ca^{2+} in alpha amylase activity?

29. What is the role of DNSA in amylase activity?

30. Discuss the factors affecting amylase activity?

31. What do you mean by optimum temperature and pH?

32. What is the role of blank or control in an enzyme assay?

4

Qualitative Analysis

4.1 INTRODUCTION

Qualitative analysis of organic compounds has been an integral part of organic chemistry laboratory, at the undergraduate level, for elucidating the structures of unknown organic compounds, long before the introduction of modern spectroscopy and chromatographic techniques. Organic spectroscopy and chromatography augment knowledge of experimental chemist by providing additional information that can be interpreted logically to help accurately predict the structure for an unknown compound.

Organic chemists often need to establish the identity of an unknown compound in the following three situations:

1. Natural product chemists need to establish the identity of new compounds isolated from natural sources.

2. A forensic scientist needs to identify small quantities of samples related to crime or lawsuites.

3. Research scientists, industrial chemists, academicians and students have to analyse compounds obtained during a chemical reaction.

Furthermore, theory and practice of establishing the identity of an organic compound is also key to introduction of research methodologies in organic chemistry. In most cases, structure elucidation requires establishing the skeletal structure, nature and position of the functional group and finally the stereochemistry. Performing qualitative analysis and collating with information obtained from spectroscopy studies helps a student to apply the knowledge of organic chemistry, that he has gained by studying over two dozen texts, to logically interpret and predict the structure of the unknown compound.

4.2 STEPS INVOLVED IN QUALITATIVE ANALYSIS OF AN ORGANIC COMPOUND

The series of steps involved in establishing the structure of an unknown compound, through qualitive analysis, can be grouped as follows:

(A) Preliminary tests: These tests give information about the general and physical properties of the unknown compound as well as the presence or absence of certain class of compounds (acidic/basic/neutral). These tests can be classified into the following sub sections:

   (i) Physical state

   (ii) Colour

   (iii) Odour

   (iv) Flame test

   (v) Solubility test: Acidic, basic or neutral behaviour

   (vi) Elemental analysis

   (vii) Heating with sodalime

   (viii) Unsaturation test

   (ix) Melting point or boiling point determination

Correct identification of the compound as acidic, basic or neutral, generally helps in subsequent identification of the functional group in the unknown compound.

(B) Functional group determination: This involves carrying out some selected classifying chemical tests and recognition of the IR and NMR bands/peaks. This approach gives information about the exact nature of the functional group present in the organic compound. This information can then be used as a basis for classification of organic compounds. Classification tests along with melting/boiling point of the unknown compound helps to identify a list of 3-4 most probable compounds.

(C) Derivatization of the unknown compound: Identification of the functional group and clarity on the class of compound through preliminary tests is used to synthesize a suitable crystalline derivative for the unknown compound. The derivative preparation involves use of a concisely defined general procedure for the unknown compound. The success of this step does not only depend on the correct application of the knowledge of organic reactions to choose a particular derivative preparation method but also requires use of the right purification and isolation technique to obtain the desired compound. The derivative, upon crystallization, is subjected to melting point determination to confirm the identity of the unknown compound.

An ideal derivative should satisfy the following conditions:

- Its preparation should involve a general reaction, which could be used to derivatize all the short-listed compounds. The reactions involving rearrangement or side reactions should be avoided.
- It should be made easily, quickly and purified readily.
- It should have well defined melting point in the range 50-250°C.
- It should possess properties distinctly different from the original compound.
- It should be able to single out one compound from the list of probable compounds. This means that the melting point of this derivative for the probable compounds must differ by at least 5°C.
- For polyfunctional compounds, the functional group chosen for derivatization should be the one which gives least ambiguous reaction.

The readers are advised to consult classified table of derivatives given in Appendix I to confirm identity of unknown compound after derivative melting point is obtained.

(D) Spectroscopic identification of the unknown compound: If the NMR, UV or IR facility is available in your institution, it is always advisable to get spectra and confirm the information obtained through the chemical tests with the data obtained in the spectrum. Otherwise, the instructor may provide the spectra of the given compound from reliable resources like webbook.nist. gov, pubchem.ncbi.nlm.nih.gov etc. Comprehending the spectral data will only enhance the understanding of the student on spectroscopic techniques and will help clarify any doubts that might have arisen in his mind while doing the chemical tests.

Note: Conventional method for qualitative organic analysis in spite of being simple has many drawbacks which need to be addressed in keeping pace with the fundamentals of green chemistry. Use of hazardous chemicals and dumping of chemicals down the drain, wastage of huge quantities of chemicals and chemical exposure are some of the key problems which require to be addressed in a logical manner. We propose the following solutions to these impending problems:

1. Explore use of newer and safer reagents for qualitative analysis. Greener alternatives have been given in this chapter wherever information was available.
2. Consult MSDS (Material Safety Data Sheet, pubchem.ncbi.nlm.nih.gov) and consult our book Volume 1, chapter 1, for details on lab safety of all the involved chemicals and their disposal before use.

3. Use proper clean up strategies. We have tried to give safe disposal methods wherever possible in this chapter.

4. Adopt semi-micro analysis and spot tests. We have highlighted the spot tests and semi-micro analysis, wherever possible, throughout this chapter. The procedure listed in this chapter involves preparations and chemical tests on semi-micro scale.

(E) Clean-up and waste disposal: As stated earlier, dumping of waste down the drain is a big issue in most colleges across India. Many instructors are not trained to propose elaborate waste disposal strategies as waste treatment/disposal in most research institutes across the world is either in-house or is done by a licensed agency. In this chapter therefore, we have attempted to provide spot tests which ensure minimum waste generation, some clean-up strategies have also been provided, wherever possible. The instructors are advised to provide clean-up strategies to students along with their instructions and ensure that separate containers are available for disposing sharp objects, halogenated hydrocarbons, other organic and toxic substances, in case an external agency has been hired for the disposal. The vessels used for waste disposal should be labelled according to their use as solids/liquids and organic/inorganic compounds. Simple garbage bins should also be available for all other types of waste. If possible, in-house paper recycle facility may be created to recycle all types of paper waste generated in the institution.

4.3   HOW TO APPROACH QUALITATIVE ANALYSIS

Identification of an unknown organic compound is more like solving a jigsaw puzzle. It is important that you keep an open mind and work in a logical, systematic and unbiased manner to test, make and reject assumptions. Working in this manner, you will successfully establish the identity of unknown compound and learning would be much more fun. Remember that your job is to discard every possibility except the correct one.

This chapter organized in a sequence of sections and sub-sections, gives an account of most common reactions for each class of organic compound with a view of identification and characterization of the unknown compound. The first *section (4.3.1)*, consisting of nine sub-sections, gives general information about the physical properties, chemical constituents and physical constants of the unknown compound. The next *section (4.3.2)*, consisting of six sub-sections, deals with each category of functional group separately. Each of these six sub-sections give information about the common reactions that can be used for identification of a functional group (**group tests**) and certain special reactions which members of each category will show (**confirmatory tests**). The confirmatory tests in each sub-section are followed by spot tests, used to characterize the members of that group. The instructors may note that spectroscopy has not been separately taken up in this book. However, the characteristic IR and NMR peaks for each functional group have been discussed at the end of each sub-section. The derivative preparation for each functional group has been taken up in the end of each sub-section. A classified table for some selected members of each category of functional group is given in Appendix I.

Note

- The students are advised to consult the classified table of organic compounds (Appendix I) for preparation of a list of probable compounds after the preliminary tests, functional group tests and spectral analysis. Final confirmation on the unknown compound may be obtained by consulting the classified tables after taking melting point of the crystallized derivative.

- Collating huge amount of information gained from doing the chemical tests can be sometimes tedious. It is therefore important that the data is recorded and analysed in a logical and systematic manner in order to arrive at the correct conclusion.

Finally, before undertaking qualitative analysis, one should ensure that the compound is pure, otherwise results may not be reliable. Instructors following this book may note that only few selected and common tests of each group have been taken up in this chapter. It is strongly felt that the students should get a good grasp on the theory and practice of doing qualitative analysis before they advance to a wider list and more complicated schemes for identification of organic compound.

4.3.1 Preliminary Tests

4.3.1.1 Physical State

Note whether the unknown organic compound is a solid or liquid. Though knowing the physical state of the compound may not give you any significant information, the knowledge that the compound is a solid and not liquid at room temperature narrows down the search substantially. For example, a liquid compound rules out carbohydrates, polyhydroxyphenols (resorcinol and hydroquinone, etc.), nitrophenols, higher acids (benzoic acid), dicarboxylic acid (oxalic acid, malonic acid), amides, polynitro compounds (*m*-dinitrobenzene), polynuclear hydrocarbons (naphthalene and anthracene), etc. Similarly, a solid compound rules out the presence of benzenes, various substituted benzenes, many aldehydes and ketones, most esters, most common alcohols, most ethers and aryl halides.

Physical state of the compound is also helpful in locating the compound in text books which arrange the compounds based on their boiling point/melting point and functional group. The same policy has also been followed in this book.

4.3.1.2 Colour

Note the colour of the unknown compound. Presence of chromophoric and auxochromic groups in a molecule usually imparts colour to most organic compounds. For example, substituted aromatic nitro compounds, substituted aromatic amines, quinones and many compounds with extended conjugated systems (like most organic dyes) are coloured.

Certain organic compounds are however coloured due to the presence of impurities. In most cases, presence of small amount of oxidation product, obtained by aerial oxidation, imparts colour to the otherwise colourless sample. For example, freshly distilled aniline is colourless, however on standing it becomes reddish brown due to aerial oxidation.

If the given unknown organic compound is pure and is a colorless liquid or a white coloured crystalline solid, it clearly rules out presence of chromophoric functional groups (example: nitro, azo etc.) as also the groups which would become chromophoric upon oxidation. For example, aromatic amino group (Please see Table 4.1).

Table 4.1   Colour and Possible Functionalities of Compound

S. No.	Colour	Possible functionalities
1.	Colourless	Simple carboxylic acid, carbohydrates, aldehydes, ketones, alcohols, ether, esters, simple amides and simple hydrocarbons, polynuclear hydrocarbons and simple halohydrocarbons.
2.	Become brown or pink on standing	Aromatic amines, diamines, phenols and aminophenols
3.	Yellow	Polynitro compounds, iodoform, nitrophenols, quinones, etc.
4.	Yellow-Orange	Nitroamines

4.3.1.3 Odour

Odour of a compound is related to its volatile nature. A more volatile compound usually has strong odour. Examining the odour of an organic compound by direct inhalation is strongly prohibited as most organic compounds have some degree of toxicity. It is advisable in all cases to refer to MSDS of every organic compound before handling. As the student gains experience, the odour of many organic compounds will become apparent during the course of regular handling. When that happens, you should make note of it. Please see Table 4.2 for identifying compounds based on their odour.

Table 4.2 Odour and Classes of Compounds

S. No.	Odour	Name/Class of compounds
1.	Fishy	Amines
2.	Rotten egg	Thiols and thioethers
3.	Vinegar	Acetic acid
4.	Fruity	Esters, Alcohols
5.	Moth Balls	Naphthalene
6.	Bitter Almond	Benzaldehyde, nitrobenzene and benzonitrile
7.	Cinnamon	Cinnamaldehyde
8.	Distinct Noxious	Low molecular weight acids

Note: Do not rely much on the odour of the compound as chemically distinct compounds may give similar odour.

4.3.1.4 Flame Test

Many organic compounds burn with a distinguishing flame and this can be used as a basis to classify them as aliphatic or aromatic. A number of exceptions to this rule, however, exist. An aromatic compound, for example, burns with a yellow, sooty flame without exception (this happens due to its relatively high carbon to hydrogen ratio). The unburnt carbon bestows black soot and yellow colour to the flame. Most aliphatic compounds on the other hand burn with yellow and non-sooty flame (due to low carbon to hydrogen ratio). As the percentage of oxygen in the organic compound increases, the flame becomes more and more clear (blue).

Note

- If the compound is flammable, it must be handled carefully in subsequent manipulations.
- Certain aliphatic compounds like carbon tetrachloride, chloroform and acetylene also give sooty flame due to high carbon to hydrogen ratio.
- Smell of burning sugar often indicates that the compound is a carbohydrate or polyhydroxy acid like tartaric acid or citric acid.

Procedure: Heat on a low flame, using nickel spatula, one drop or 10 mg of the unknown compound. Raise the temperature gradually until the compound is completely burnt. Observe any characteristic odour, soot and colour of the flame when the compound burns.

Note

- Do not remove the spatula from the burner until the compound has completely charred/oxidized.

- Avoid taking too much of compound as you may unknowingly fill your lab with harmful fumes.

4.3.1.5 Solubility Test

Solubility tests should be carried out on every unknown compound as they are rapid and consistent. They give preliminary information about the structural composition of organic compound. In general, if the given organic compound (0.1 g or 0.2 mL) forms homogenous mixture with solvent (3.0 mL; 3-5% solution w/v or v/v), it is considered to be soluble, as per our generalisation. The solubility tests can be useful in the following three ways:

- To classify the compounds as acidic, basic and neutral, except for cold water soluble compounds.

- Sometimes to classify the compounds into specific functional groups. (See Table 4.3) For example, carboxylic acids are soluble in both 10% aqueous NaOH and 5% aqueous $NaHCO_3$ whereas phenols are usually soluble in former and not in latter, due to weakly acidic nature.

- To separate a mixture of organic compounds based on their solubility.

Five commonly used reagents for the solubility tests are:

(1) Water, (2) 10% aqueous NaOH (3) 5% aqueous $NaHCO_3$, (4) 5% aqueous HCl, and (5) cold concentrated (96%) H_2SO_4.

Note

- Solids take slightly longer time to dissolve than liquids. It is advisable to powder solids well before dissolution.

- A coloured compound on dissolution will lead to coloured solution.

- If the given compound is soluble in cold water, it is likely to be soluble in 10% aqueous NaOH, 5% aqueous $NaHCO_3$ and 5% aqueous HCl as all these are made in water. Therefore, doing solubility tests for such a compound in solvents other than water will not provide any additional information about its acidic or basic character.

- You must always look for increased solubility of your compound in cold acid or base as compared to water before making any conclusion about its acidic or basic character.

- It is advisable also that you check the pH of the aqueous solution (*use pH paper*) to gain information about the acidic/basic character of the compound.

- Do not rely on just the visual tests for solubility, in case of carboxylic acids/phenols (acidic compounds) or amines (basic compounds). It is always advisable to use the supernatant liquid obtained after dissolution to precipitate back the dissolved compounds. This can be done by adding mineral acid in case of carboxylic acid/phenol and alkali in case of amines. Any fresh precipitate will confirm that the compound was soluble in the original solution.

- Similarly, if the given compound is basic (see HCl solubility), it will have to be certainly an amine. (please correlate with your extra element test and see Table 4.3).

- Tap the test tube vigorously with your finger after you have added the solvent, this will ensure thorough mixing of the contents.

- In case of colurless liquids, sometimes two colorless phases (water and the organic compound) lie one below the other and may be mistaken as one phase. This error can usually be avoided by shaking the test tube vigorously. The solution immediately becomes cloudy if two phases are present.

- In case the unknown liquid does not dissolve, as you continue tapping, the tiny drops of the unknown liquid will coalesce into a larger drop either at the bottom or the top of the solvent, depending on the relative densities of the unknown liquid and the solvent.

- Mixtures in sulphuric acid may require vigorous stirring with a glass rod to ensure thorough mixing of the solute and solvent, because sulphuric acid is very viscous and dense.

- Weakly basic compounds like alcohols, aldehydes, ketones, and esters, are soluble in strong acids like concentrated sulphuric acid due to formation of oxonium salts. Alkenes dissolve due to formation of hydrogensulphates.

- Many compounds containing N & S are neutral to due to weakly basic nature. They have unshared electron pair, hence will dissolve in concentrated acids like concentrated sulphuric acid. However, no additional information for such cases is obtained by doing sulphuric acid solubility. Therefore, this test may be generally avoided.

Table 4.3 gives details on the solubility of various class of functional groups in all the five types of solvents discussed above. Figure 4.1 uses a concept map to classify the organic compounds into various functional groups based on the solubility chart shown in Table 4.3. You may refer to either one depending upon your convenience.

1. Solubility in water: Solubility of a compound in a particular solvent depends on its structure and number of carbon atoms present. As a general rule, compounds having polar functional groups like alcohols, carboxylic acids, polyphenols, carbohydrates, amines, etc., and having up to four to five carbons are soluble in water due to formation of solute-solvent hydrogen bonds. This

Table 4.3 Solubility Chart

S. No.	Functional group	Water	10% Aq. NaOH	5% Aq. NaHCO₃	5% Aq. HCl	Conc. H₂SO₄
1.	Carboxylic acid	$-^a$	+	+	–	$+^b$
2.	Aldehydes	$-^a$	–	–	–	$+^b$
3.	Ketones	$-^a$	–	–	–	$+^b$
4.	Phenols	$-^d$	+	$-^c$	–	+
5.	Esters	–	$-^e$	–	–	+
6.	Alcohols	$-^a$	–	–	–	$+^b$
7.	Aryl halides	–	–	–	–	–
8.	Amines	–	–	–	+	F
9.	Amide	–	$-^e$	–	–	F
10.	Nitro	–	–	–	–	F

a. except few small chain compounds **b.** soluble due to oxonium salt formation. **c.** except for few nitro phenols **d.** except for polyphenolic compounds. **e.** Except when boiled. **f.** Avoid this test for N-containing compounds as the reaction can be violent

is because as the carbon chain length increases, the non-polar carbon chain tends to dictate the solubility of the compound. It is, therefore, important to interpret the solubility of the compound judiciously keeping all the factors listed previously in mind. (See Table 4.3 for details)

Procedure: Take solid (0.1 g)/liquid (0.2 mL) in a test tube, add water (1.0 mL) and shake. Follow this by adding more water (2.0 mL) and shake again. If the compound is soluble, take a dry glass rod and dip into the aqueous solution and introduce the solution on the pH paper to check the pH of the solution.

If however the compound is insoluble in water, warm the solution gently and see if the compound dissolves. Note all observations carefully.

2. Solubility in 10% aqueous NaOH: Water insoluble organic acids (pK$_a$ 4-5) and phenols (pK$_a$ = 10) dissolve in 10% NaOH (pH>14) solution due to formation of soluble carboxylate or phenoxide salts.

$$RCOOH + OH^- \rightleftharpoons RCOO^- + H_2O$$

$$\text{(phenol)} + OH^- \rightleftharpoons \text{(phenoxide)} + H_2O$$

$$RCOOR' + OH^- \xrightarrow{\text{Heat}} RCOO^- + R'OH$$

$$RCONH_2 + OH^- \xrightarrow{\text{Heat}} RCOO^- + NH_3$$

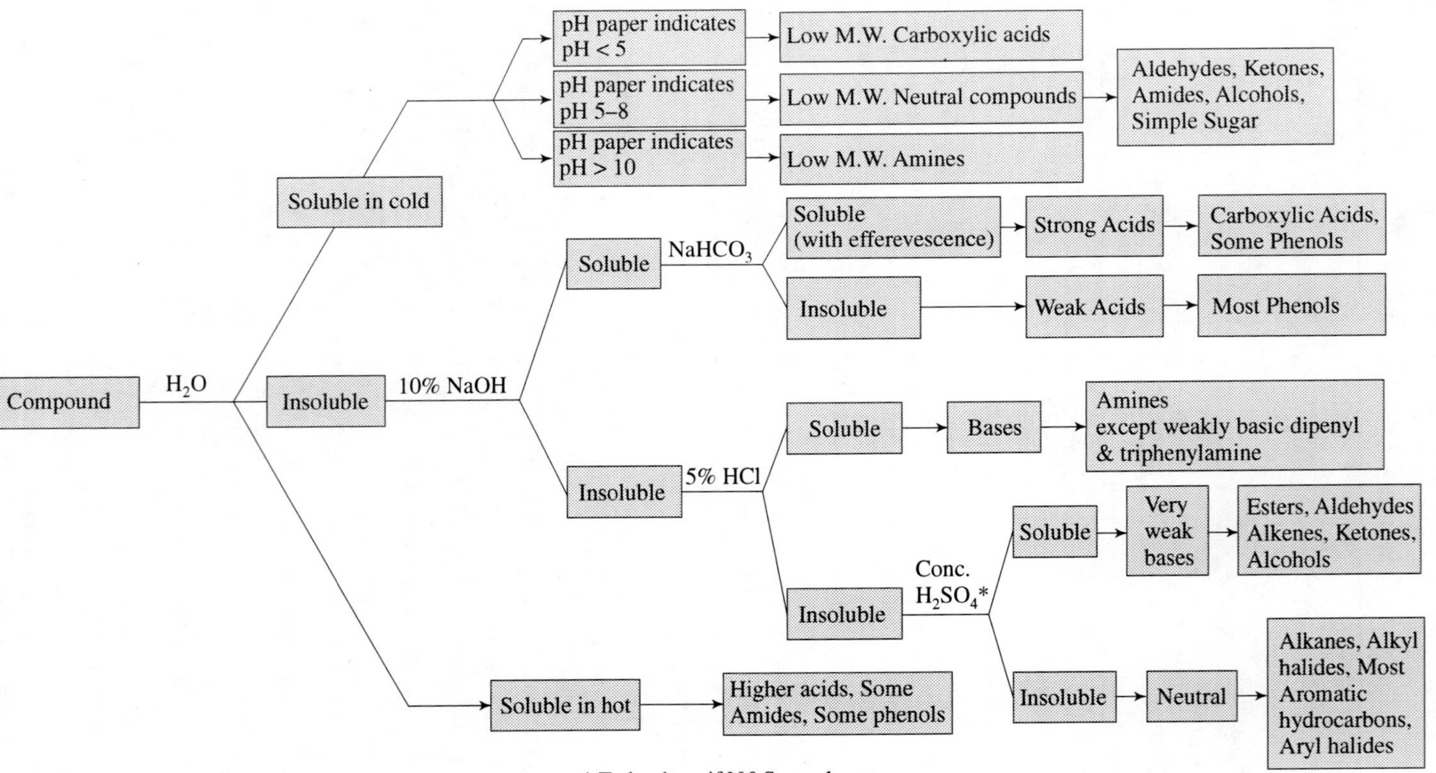

Fig. 4.1 Concept Map for Solubility of Organic Compounds

The hydrolysis of amides and esters is strongly supported by 10% sodium hydroxide solution however the process is too slow at room temperature to allow an amide or ester to dissolve in 10% NaOH within a few minutes. Heating the mixture however may produce, soluble carboxylate salt, due to hydrolysis of the amide or ester. Thus for solubility classification, it is advisable to carry out the test in cold 10% aqueous sodium hydroxide solution only.

Procedure: Take solid (0.1 g)/liquid (0.2 mL) in a test tube, add 10% aqueous sodium hydroxide solution (1.0 mL) and shake. Follow this by adding more NaOH solution (2.0 mL) and shake again. If the compound appears insoluble, take the supernatant liquid and add 5% aqueous hydrochloric acid dropwise until the medium is acidic. Appearance of fresh precipitate or turbidity confirms that the compound is a carboxylic acid, phenol, aliphatic primary or secondary nitro compounds. Correlate with the results of extra element test before arriving at any conclusion.

Note

- This test is performed for water insoluble compounds only.
- Some substituted phenols are insoluble due to precipitation of sodium salts. They, however, dissolve on dilution with water followed by warming.

3. Solubility in 5% aqueous sodium bicarbonate: When the given organic compound is insoluble in water but soluble in 10% aqueous NaOH, then solubility in 5% aqueous sodium bicarbonate is used to distinguish between a carboxylic acid or a sulphonic acid and a phenol. 5% Aqueous sodium bicarbonate solution is weakly basic with a pH of ~9. It dissolves most water insoluble carboxylic acids or sulphonic acids ($pK_a < 7.5$) by converting them to a water-soluble salt. Brisk effervescence due to evolution of carbon dioxide is characteristic of carboxylic acids when dissolved in sodium bicarbonate solution. Most phenols, except a few nitrophenols like 2,4-nitrophenol, picric acid, etc., do not dissolve in aqueous sodium bicarbonate and also do not produce effervescence. This is due to their weakly acidic nature. Water soluble phenols however dissolve in 5% aqueous bicarbonate solution without liberating carbon dioxide.

$$RCOOH + HCO_3^- \longrightarrow RCOO^- + H_2O + CO_2$$

Procedure: Take solid (0.1 g)/liquid (0.2 mL) in a test tube, add 5% aqueous sodium bicarbonate solution (1.0 mL) and shake. Follow this by adding another 2.0 mL of the solution and shake again. See if the compound dissolves. Look for effervescence, if the compound dissolves with effervescence it indicates presence of carboxylic acid or sulphonic acid or nitrophenol. Correlate with extra element test and functional group test, before concluding.

Note

- Primary and secondary aliphatic nitro compounds do not dissolve due to weak acidic character.
- Amino acids evolve carbon dioxide after standing for some time.
- This test should be performed for water-insoluble compounds only.
- During low temperatures, solubility of aromatic acids in aqueous sodium bicarbonate may be greatly diminished. In such cases, prepare aqueous solution of sodium bicarbonate in warm (not boiling) water and proceed as above.

4. Solubility in 5% aqueous hydrochloric acid: Water-insoluble organic compounds, which dissolve in 5% aqueous HCl, contain a basic nitrogen functionality in the molecule (primary, secondary, tertiary aliphatic amines, aryl-alkyl amines and anilines). Nearly all types of amines (except for a few like diphenyl and triphenyl amines) react with HCl to produce ammonium salts, which are usually soluble in water.

$$RNH_2 + HCl \longrightarrow RNH_3^+Cl^-$$

Procedure: Take solid (0.1 g)/liquid (0.2 mL) in a test tube, add 5% aqueous hydrochloric acid (1.0 mL) and shake. Follow this by adding another 2.0 mL of the acid and shake again. See if the compound dissolves. Take the supernatant liquid in a fresh test tube and make it alkaline by adding 10% aqueous NaOH solution, dropwise. Fresh precipitate or reappearance of oily droplet (in case of liquid compound) indicates the presence of amine.

Note

- Aromatic amines having more than one benzene ring like diphenylamine and triphenylamine are insoluble in HCl due to decreased basic character.
- Ammonium salts precipitate in the presence of excess of acid. But, they dissolve in solution again on dilution with water. A positive test with alcoholic silver nitrate also confirms ammonium salt.
- This test should be performed for water-insoluble compounds only.

5. Solubility in concentrated sulphuric acid (96%):   All neutral, unknown organic compounds containing nitrogen and/or oxygen dissolve in concentrated sulphuric acid because of protonation to form ionic salts. Highly polar nature of sulphuric acid is responsible for the dissolution of these protonated compounds. This dissolution process may be accompanied by production of large quantities of heat and/or may lead to change in the colour of the mixture. Organic compounds containing nitrogen or oxygen functionalities in the molecule (except amines and carboxylic acids) are weak bases. In dilute solutions of mineral acids,

$$\underset{H_3C}{\overset{H_3C}{>}}\!\!=\!CH_2 + H_2SO_4 \longrightarrow \underset{H_3C}{\overset{H_2C}{\underset{H_3C}{>}}}\!\overset{\oplus}{C}-CH_3 \longrightarrow \cdots$$

Polymer

moderate concentrations of conjugate acids of these compounds are present, hence they are insoluble. In concentrated sulphuric acid, these protonated species (or conjugate acids) are present in significantly higher concentrations, due to high stability of the conjugate acid form. This is responsible for their solubility in concentrated sulphuric acid as compared to 5% aqueous HCl.

$$ROH + H_2SO_4 \rightleftharpoons ROH_2^+ + HSO_4^-$$

$$\underset{R}{\overset{O}{\underset{}{C}}}\!\!_R + H_2SO_4 \rightleftharpoons \underset{R}{\overset{\overset{+}{O}H}{\underset{}{C}}}\!\!_R + HSO_4^-$$

$$RC \equiv N + H_2SO_4 \rightleftharpoons RC \equiv NH^+ + HSO_4^-$$

The alkenes dissolve in sulphuric acid due to protonation of the vinylic carbon to form carbocations. The carbocations may further react with bisulphate ion, giving alkyl hydrogen sulphates, both carbocation and bisulphates are soluble in sulphuric acid. The carbocations may also add on to unprotonated alkene to give insoluble brown polymers.

Polynuclear hydrocarbons like anthracene undergo sulphonation with concentrated sulphuric acid to produce darkening. Benzene and substituted benzenes like chloro and bromo derivatives do not undergo sulphonation at room temperature and are insoluble in sulphuric acid. Presence of two or more alkyl groups on the aromatic ring permits the compound to be sulfonated easily at room temperature. Therefore, polyalkylbenzenes such as 1,2,3,5-tetramethylbenzene (isodurene), 1,3,5-trimethylbenzene (mesitylene) and alkyl phenyl ethers dissolve rather readily in sulphuric acid due to sulphonation.

$$ArH + 2H_2SO_4 \longrightarrow ArSO_3H + HSO_4^- + H_3O^+$$

Procedure: Take sulphuric acid (3.0 mL) in a dry test tube, add solid (0.1 g) or liquid (0.2 mL) and shake vigorously. See if the compound dissolves. Look for any changes in colour, evolution of heat, charring or formation of insoluble polymer.

Note

- For solubility classification purpose, compounds producing heat and/or color changes or black tarry substance on adding concentrated sulphuric acid should be classified as soluble, even if the sample does not apparently dissolve. For example, carbohydrates darken with evolution of CO_2 or CO gas.

- It is possible that some compound reacts with sulphuric acid to produce a new compound which is insoluble in sulphuric acid. So, look for any kind of change on dissolution before arriving at a conclusion.

- This test should only be performed for water insoluble neutral compounds lacking N or S. Avoid treating water-insoluble compounds, thought to be amines, with concentrated H_2SO_4 because the heat produced may lead to a violent reaction.

- Concentrated sulphuric acid is highly corrosive and may lead to severe burns. Hence, should be handled with care.

Safe Waste Disposal

- Pool all aqueous test solutions (except that containing sulphuric acid) in a beaker and adjust the pH to 7 by dropwise adding 5% aqueous HCl or 5% aqueous NaOH (as applicable), with stirring. Use pH paper to check the pH. Transfer this neutral solution to the separating funnel and add ether (4-5 mL). Stopper the funnel and shake it to mix the layers. Allow the layers to separate, and remove the lower aqueous phase. Run the aqueous phase down the sink or place it in the container for aqueous inorganic waste. Place the ether solution in the flammable (organic) waste container if you have determined that no halogen is present or in the halogenated waste container if you have found that your unknown compound contains a halogen.

- Cautiously transfer the concentrated sulphuric acid test solution into a beaker containing water (30 mL) (do not attempt the reverse). Dropwise add aqueous solution of sodium carbonate till neutral to pH paper (caution: foaming!). Add some crushed ice to cool the solution to room temperature (if required). Transfer this neutral solution to a separating funnel and extract it with ether (4-5 mL, two times). Then proceed as mentioned above for ethereal layer.

4.3.1.6 Elemental Analysis

Organic chemists usually do not perform chemical tests to detect the presence of carbon, hydrogen and oxygen. This is because the organic compounds invariably have carbon and hydrogen, barring few exceptions like carbon tetrachloride. On the other hand, oxygen is difficult to detect. Detection of other commonly occurring elements like nitrogen, halogens and sulphur is valuable as it helps to characterize the organic compounds into specific functional types. For example, nitrogen containing, sulphur containing, simple hydrocarbon, etc. These elements are therefore called **extra elements**. Detection of extra elements is usually done by chemical tests involving reaction with aqueous reagents which are usually immiscible with organic compounds. (Remember, organic compounds are relatively non-polar and contain covalent bonds which are difficult to break at room temperature) It is therefore desirable to perform the sodium fusion test with organic compounds and convert the extra elements to water soluble sodium salts like cyanide, sulphide and halides, prior to performing elemental analysis. This aqueous solution called **Lassaigne's extract** (LE), is used to perform individual tests for extra elements.

$$\text{C, H, O, N, S, X} \xrightarrow{\ \text{Na}\ } \begin{cases} \text{NaX} \\ \text{Na}_2\text{S} \\ \text{NaOH} \\ \text{NaCN} \end{cases}$$

$$X = \text{Halogen}$$

In case both nitrogen and sulphur are present in an organic compound, it is possible that a mixture of sodium sulphide and sodium cyanide is obtained. In some cases however, sodium thiocyanate may be obtained, which decomposes in excess of sodium to sodium sulphide and sodium cyanide:

$$\text{Na} + \text{S} + \text{C} + \text{N} \longrightarrow \text{NaSCN}$$

$$\text{NaSCN} + 2\,\text{Na} \longrightarrow \text{Na}_2\text{S} + \text{NaCN}$$

Note

- Wear proper personal protective equipment (PPE) while performing these tests.
- Blank tests should be performed whenever in doubt.
- The extra element tests are rather sensitive, so all aqueous solutions should be prepared in clean test tubes using double distilled or deionized water only.
- Compounds that are known to react explosively with molten sodium are nitro alkanes, organic azides, diazo esters, diazonium salts and some

aliphatic polyhalides such as chloroform or carbon tetrachloride. These should be analyzed by a small-scale procedure.

Elemental analysis of unknown compound is beneficial in the following ways:

- The knowledge of extra elements assists in the correct selection of classification experiments like tests for functional groups (see *section 4.3.2*).
- Helps in the identification of relevant procedures for the derivative preparation (see *section 4.3.2*).
- Helps in the correct interpretation of the spectroscopic data for identification of the structure of unknown compounds.

Procedure: Take out sodium metal from the bottle of kerosene. Wipe off all kerosene with filter paper. Hold sodium with a pair of forceps (caution!) and use a clean and dry knife to cut a small piece of clean and shiny sodium metal, about 3-4 mm long. Take a clean and dry test tube (approximately 50 × 13 mm)/ignition tube made of soft glass (not Pyrex or Kimax). Hold it firmly with the help of a test tube holder and place the freshly cut piece of sodium in the test tube. Heat the lower part of the test tube/ignition tube until the sodium melts to form a shiny globule and sodium vapours begin to rise in the test tube, remove the test tube from flame, cool for thirty seconds and add solid compound (50.0 mg)/ liquid compound (3 drops). Heat the test tube first gently and when the fumes begin to subside, heat vigorously until the bottom of the test tube becomes red hot. Continue heating for another 2-3 minutes to ensure complete reaction of the compound with sodium metal. Plunge the red-hot tube immediately in a clean porcelain dish containing ~10.0 mL of distilled water. A vigorous reaction occurs as the tube breaks and the excess of sodium reacts with water. Hence, it is important to cover the porcelain dish with wire gauze while plunging the test tube in it. Repeat two more fusions and plunge in the same porcelain dish. Break the broken pieces of glass with a glass rod and boil the contents of porcelain dish to about half its volume. Filter the clear solution. If the filtrate is dark colored, it indicates decomposition is not complete. Repeat the entire fusion process again and use the clear solution for performing individual tests for extra elements. This clear filtrate is called **"Lassaigne's extract"** or **"sodium fusion extract"**.

Note

- Always wear personal protective equipment (PPE) while doing any experiment in the lab. (Refer chapter 1, volume 1).
- First time preparation of the fusion extract should be done under the supervision of instructor.
- Never handle sodium metal with bare hands. Always use foreceps.

- Sodium is a very reactive metal and it reacts violently with water and oxygen both. Therefore, it is stored under unreactive solvent like kerosene.

- Sodium piece should be free from oxide and hydroxide coating (ensure that you take a shiny part by scrapping the yellow coatings with the help of a knife).

- Take freshly cut piece of sodium in one ignition tube at a time.

- In case of volatile liquids, it is advisable to use a paste of compound with sodium carbonate, to avoid loss by volatilization.

- In case of liquid compounds, heating with sodium metal should be continued until all the oily droplets in the ignition tube have disappeared. This will ensure complete decomposition of the compound.

- Ensure that the Lassaigne's extract is alkaline before performing the elemental analysis. If not, add 2-3 drops of sodium hydroxide solution, before proceeding.

- In case the ignition tube breaks while doing the fusion, dispense it in a beaker containing small amount of ethanol. This will ensure the complete decomposition of the unused sodium by ethanol.

- In case the compound reacts violently with sodium, it is advisable to remove it from the flame and reheat again after the reaction subsides.

Safe Waste Disposal

- After completion of elemental analysis, the unused sodium fusion test solutions should be transferred to the aqueous solution container for disposal.

4.3.1.6.1 *Test for nitrogen*

1. Lassaigne's Sodium Test: This is the most commonly used test to identify presence of nitrogen in an organic compound. The organic compound is fused with molten sodium. Carbon, nitrogen and oxygen present in the compound form alkaline Lassaigne's extract containing sodium cyanide. A few crystals of ferrous sulphate are then added to Lassaigne's extract. The solution immediately turns greenish-grey due to the hydroxides of iron. Boiling the alkaline mixture leads to partial aerial oxidation of ferrous ions to ferric ions. Acidification of cooled solution with dilute sulphuric acid dissolves hydroxides of ferrous and ferric ions. A prussian blue coloured precipitate is obtained due to formation of ferriferrocyanide, commonly known as **Prussian blue.**

Procedure: Take Lassaigne's extract (2-3 mL) in a pyrex glass test tube and add ferrous sulphate (0.1-0.2 g). An intense greenish-grey* precipitate of ferrous hydroxide appears immediately. If the solution remains clear, add 2-3 drops of sodium hydroxide solution. Now boil the mixture gently for about five minutes. This ensures partial oxidation of ferrous ions to ferric ions. Cool and acidify the solution with dilute sulphuric acid, to ensure complete dissolution of hydroxides of iron. A prussian blue colouration or precipitate indicates presence of nitrogen.

* If sulphur is present in the unknown compound, a black precipitate of ferrous sulphide may be observed at this stage.

$$FeSO_4 + 2NaOH \longrightarrow Na_2SO_4 + Fe(OH)_2$$

$$6NaCN + Fe(OH)_2 \longrightarrow Na_4[Fe(CN)_6] + 2NaOH$$

$$Fe^{2+} \xrightarrow[\text{Warm}]{[O]/Air} Fe^{3+}$$

$$3Na_4[Fe(CN)_6] + 2Fe_2(SO_4)_3 \longrightarrow Fe_4[Fe(CN)_6]_3 + 6Na_2SO_4$$
$$\textbf{Prussian blue}$$

Note

- In case both nitrogen and sulphur are present, on acidification crystals of ferrous sulphide dissolve along with hydroxides of iron, leaving the Prussian blue colouration/precipitate.

- A faint green colouration upon acidification usually indicates incomplete reaction with sodium.

- It is not advisable to add ferric chloride (as source of ferric ions) during the nitrogen test, as yellow ferric chloride makes the prussian blue colour appear green.

- Only dilute sulphuric acid should be used for acidification as nitric acid will oxidize all ferrous ions and no prussian blue will be formed whereas hydrochloric acid will form yellow ferric chloride making blue colour appear green.

- Lassaigne's sodium test is a test for nitrogen in the presence of carbon. If you wish to detect nitrogen in some inorganic compounds like hydrazine, then you should add a non-volatile nitrogen free organic compound like sucrose or glucose before performing the fusion test.

- In the absence of nitrogen, the acidified solution will have pale yellow colour of iron salts.

2. Ammonium polysulphide test for nitrogen: This test can be used in addition to sodium fusion test to confirm the presence of nitrogen in the given organic compound. The cyanide ions are converted into thiocyanate ions on heating fusion extract to dryness with yellow ammonium polysulphide. The thiocyanate ions react with ferric chloride to give blood red coloration due to formation of iron-thiocyanate complex.

Procedure: To the fusion extract (2.0 mL) placed in a clean test tube, add 10% yellow ammonium polysulphide solution (2 drops). Evaporate the mixture on hot water bath to complete dryness. Add 5% aqueous solution of hydrochloric acid (5.0 mL) to the residue and warm. Filter the solution. To the filtrate add 5% aqueous ferric chloride solution (3 drops). Appearance of red colour indicates the presence nitrogen.

$$NaCN + \underset{\text{Ammonium sulphide}}{(NH_4)_2S_x} \longrightarrow NaSCN + (NH_4)_2S_{x-1}$$

$$6NaSCN + FeCl_3 \longrightarrow \underset{\text{Red}}{Na_3Fe(SCN)_6} + 3NaCl$$

Safe Waste Disposal

- Place all the solutions in the aqueous solution container.

4.3.1.6.2 Test for sulphur

Sulphur in the organic compound is converted to sulphides on fusion with sodium. It can be identified in the aqueous solution by either doing the lead acetate test or sodium nitroprusside test. In the lead acetate test, the fusion extract is acidified with acetic acid and then 1% solution of lead acetate is added. A black precipitate of lead sulphide indicates sulphur.

In the Lassaigne's sodium test, a few drops of freshly prepared sodium nitroprusside are added to the sodium fusion extract. Violet colour resembling potassium permanganate solution confirms the presence of sulphur.

1. Lead acetate test: Take Lassaigne's extract (2.0 mL) in a test tube, acidify with acetic acid and add 1% lead acetate solution (4-5 drops). A black precipitate of lead sulphide indicates sulphur.

$$Na_2S + \underset{\text{Lead acetate}}{Pb(CH_3COO)_2} \longrightarrow \underset{\text{Black ppt.}}{PbS\downarrow} + 2CH_3COONa$$

2. Lassaigne's Sodium Test (Sodium Nitroprusside Test): Take Lassaigne's extract (2.0 mL) in a test tube and add 0.1% freshly prepared solution of sodium nitroprusside (3-4 drops). A bright purple coloration (resembling permanganate solution) indicates sulphur. The coloration slowly fades on standing.

$$Na_2S + \underset{\text{Sodium nitroprusside}}{Na_2[Fe(CN)_5NO]} \longrightarrow \underset{\text{Purple}}{Na_4[Fe(CN)_5NOS]}$$

Note

- *While detecting sulphur using lead acetate test:* Check the acidity with litmus paper before adding lead acetate. Only acetic acid is used for acidification as sodium acetate formed is soluble in water and does not interfere with the test. Do not use strong mineral acids like sulphuric acid for acidification. If sulphuric acid is used, it will form white precipitate of lead sulphate on reaction with sodium sulphide.

- *While detecting sulphur using sodium nitroprusside test*: Aqueous solution of sodium nitroprusside is unstable and therefore a freshly prepared solution should be used on each occasion.
- Sodium nitroprusside test for sulphide is very sensitive. Therefore, it is essential that the test tubes used should be absolutely clean.

Safe Waste Disposal

- Place all the solutions obtained from either test for sulphur in the aqueous solution container.

4.3.1.6.3   *Test for nitrogen and sulphur*

As stated earlier, it is important to use excess of sodium for the sodium fusion test otherwise sulphur and nitrogen present together in the compound may form sodium thiocyanate, NaSCN, which will give negative test for nitrogen and sulphur both. Presence of thiocyanate can however be confirmed by acidifying the Lassaigne's extract with hydrochloric acid and then adding a few drops of ferric chloride solution. Blood red colouration of sodium hexathiocyanato ferrate(III) confirms the presence of both nitrogen and sulphur.

In case excess of sodium is present in the fusion extract, sodium thiocyanate is decomposed to sodium sulphide and sodium cyanide as shown earlier (See page 101). In such cases the fusion extract will respond to test for nitrogen and sulphur, while the thiocyanate test will be negative.

$$6\,NaSCN + FeCl_3 \longrightarrow Na_3[Fe(SCN)_6] + 3\,NaCl$$
Blood red colour

Procedure: Acidify the fusion extract (1.0 mL) with dilute hydrochloric acid and add ferric chloride solution (2-3 drops). Appearance of blood red colouration indicates nitrogen and sulphur are present.

Note

- Ensure that the medium is acidic before adding ferric chloride. This will ensure that brown precipitate of ferric hydroxide does not interfere with the test.

- In case Lassaigne's sodium tests for nitrogen and sulphur are positive, you may avoid doing the combination test. However, when both nitrogen and sulphur tests are negative, it is important that the combination test is performed to confirm the presence of thiocyanate ion.

4.3.1.6.4   *Test for halogens*

1. Beilstein test: Copper or copper oxides decompose organo-halogen compounds to produce copper halides, which give green colouration in the flame, due to volatile nature. This fact can be used to detect the presence of halogen in the given organic compound. This test is performed prior to the sodium fusion test for halogens.

Procedure: Take a copper wire and roll it into a loop. Fit this copper wire in a cork and hold it by the cork. Now, heat the looped end of the copper wire until any green colouration, due to impurities on copper wire, disappear completely. Dip the copper loop into the organic compound (solid/liquid) and introduce to the edge of the flame. Fresh green colouration on heating indicates that halogens may be present.

Note

- Beilstein test is a more sensitive test than Lassaigne's sodium test for halogens, as it is unaffected by the presence of nitrogen or sulphur.
- This test may fail in case of very volatile compounds which might evaporate before decomposition by copper.
- A false green flame may be produced due to traces of halogen containing impurities on copper wire. It should therefore be heated properly before testing the unknown compound for halogen.
- Fluoro compounds do not give this test as copper fluoride is not volatile.
- Certain halogen-free amide compounds like urea, thiourea, cyanides and other pyridine compounds form volatile copper cyanides on heating, which also give green coloration.
- Absence of green colour is a confirmation that no halogen is present. However, a positive Beilstein test indicates probable presence of halogen, which should be confirmed by doing the Lassaigne's sodium test for halogens.
- The test only indicates presence of halogens. It cannot be used to distinguish chlorine, bromine and iodine present in the compound.

2. Lassaigne's Sodium Test (Silver Nitrate Test): *When both nitrogen and sulphur are absent:* Sodium halides formed in the sodium fusion extract react with silver nitrate to give white or yellow precipitate due to formation of silver halides. These silver halides can be distinguished on the basis of their colour and solubility in aqueous ammonia.

Silver chloride, silver bromide and silver iodide have different solubilities in aqueous ammonia (5% ammonium hydroxide solution). Silver chloride is soluble in aqueous ammonia due to the formation of soluble complex, $[Ag(NH_3)_2]Cl$. Silver bromide is partially soluble because it only partially forms the complex bromide. Silver iodide does not react with the aqueous ammonia and thus remains insoluble.

When nitrogen and/or sulphur is present: In case nitrogen and/or sulphur is present, addition of silver nitrate to the acidified Lassaigne's extract will give respectively white precipitate of silver cyanide or black precipitate of silver sulphide. In either case, the white/yellow colour of silver halide will no longer be identifiable. Therefore, cyanide and sulphide ions are removed prior to addition of silver nitrate solution. Acidified Lassaigne's extract when subjected to prolonged heating, expels cyanides and sulphides as HCN and H_2S gases. The cyanide and/or sulphide-free Lassaigne's extract is then subjected to reaction with silver nitrate solution.

$$NaCN + HNO_3 \longrightarrow HCN + NaNO_3$$

$$Na_2S + 2\,HNO_3 \longrightarrow H_2S + 2\,NaNO_3$$

$$NaCl + AgNO_3 \longrightarrow AgCl\downarrow + NaNO_3$$
$$\textbf{(White)}$$

$$NaBr + AgNO_3 \longrightarrow AgBr\downarrow + NaNO_3$$
$$\textbf{(Pale Yellow)}$$

$$NaI + AgNO_3 \longrightarrow AgI\downarrow + NaNO_3$$
$$\textbf{(Yellow)}$$

$$AgCl + 2\,NH_4OH \longrightarrow [Ag(NH_3)_2]Cl + 2H_2O$$

Procedure: Take Lassaigne's extract (2.0 mL) in a test tube and acidify with dilute nitric acid until the solution is just acidic to litmus. Then proceed as in (a) or (b), as applicable.

(a) ***In case N and/or S is present:*** Transfer the solution to a clean porcelain dish and evaporate the solution to dryness. Now, cool the contents, extract the residue with dilute nitric acid and test for halogens by adding a few drops of silver nitrate solution to one part of the extract. Note the formation of yellow/white precipitate.

(b) ***In case N/S is absent:*** Add silver nitrate (1.0 mL) directly to the acidified Lassaigne's extract. A white or yellow precipitate indicates halogen.

Decant the supernatant liquid and collect the precipitate of silver halide in (a) or (b).

Note

- Nitric acid is added to neutralize the alkaline Lassaigne's extract. This prevents formation of brown or black precipitate of silver oxide (or silver hydroxide), on addition of silver nitrate solution.

3. Silver nitrate test: Silver halides (chlorides, bromide and iodide) may be identified on the basis of the colour and solubility of their respective silver halides in aqueous ammonia. Presence of bromine or iodine may be further confirmed by doing the layer test.

Procedure: Presence of chlorine, bromine or iodine may be tested by performing the following tests sequentially after doing the step (a) or (b) for Lassaigne's sodium test (point no. 2), as applicable. Decant the precipitate of silver halide and dissolve it in aqueous ammonia and check the following:

- A white coloured precipitate of silver chloride, completely soluble in ammonia confirms the presence of chlorine. The precipitate reappears on adding dilute nitric acid.
- A pale-yellow precipitate of silver bromide partly soluble in ammonia indicates bromine
- A deep yellow precipitate insoluble in ammonia indicates iodine.

4. Layer test: Silver halides in the sodium fusion extract are oxidised by oxidising agents like nitric acid, potassium permanganate or chlorine water to the corresponding halogens (X_2). These halogens impart colour to the organic layer (carbon tetrachloride or dichloromethane). If bromine is present then a deep orange to brown colour is observed whereas a violet colour is seen if iodine is present.

$$2\,NaBr + Cl_2(H_2O) \longrightarrow 2\,NaCl + Br_2(CH_2Cl_2)$$
Orange-brown
colouration

$$2\,NaI + Cl_2(H_2O) \longrightarrow 2\,NaCl + I_2(CH_2Cl_2)$$
Violet
colouration

Procedure: Take sodium fusion extract (1.0 mL) and add concentrated nitric acid (0.5 mL) (a drop or two of potassium permanganate may also be added at this stage). Warm the solution on the Bunsen burner, but do not boil. Cool the solution, add carbon tetrachloride or dichloromethane (2 3 mL) and shake. Orange-brown colour in the organic layer indicates bromine, and a violet colour indicates iodine. If no characteristic colour appears, and a white coloured precipitate is obtained in the silver nitrate test, the halogen is chlorine.

Note

- These halogen tests are useful only when the compound contains one type of halogen atom.
- Silver fluorides are water soluble therefore cannot be identified by silver nitrate test.
- In the layer test, do not boil the solution after adding concentrated nitric acid as bromine or iodine formed by oxidation, may escape on boiling.
- A blank test for halogens with the water sample used for formation of fusion extract must be performed to rule out any anomaly.

Safe Waste Disposal

- Filter and collect all the silver and lead salts and put them in the hazardous solid waste container.
- Place the dichloromethane or carbon tetrachloride in the organo-halogen solvent container.
- Neutralize the acidic solutions with 5% aqueous sodium bicarbonate solution and the basic solutions with 5% aqueous hydrochloric acid before transferring them to the aqueous solution container.

5.   Test for presence of more than one halogen: When an organic compound contains more than one halogen, the relative rate of oxidation of silver halides becomes the basis of distinction. The rate of oxidation of sodium halides follows the order: NaI>NaBr>NaCl.

Procedure: Acidify Lassaigne's extract (2.0 mL) with glacial acetic acid. Warm, cool and add carbon tetrachloride (2.0 mL) followed by addition of sodium nitrite solution (2.0 mL, 20%), dropwise with shaking. A purple colouration in the organic layer indicates iodine. Add more sodium nitrite solution until precipitation stops. Remove the aqueous layer carefully with the help of a dropper. Introduce fresh carbon tetrachloride (2.0 mL) and shake. Continue removal of aqueous layer and addition of fresh carbon tetrachloride (2.0 mL) until the organic layer is colourless, this ensures complete removal of iodine.

$$2\,NaI + 2NaNO_2 + 4CH_3COOH \longrightarrow I_2 + 2NO + 4CH_3COONa + 2H_2O$$

Next, boil the aqueous layer to expel any nitrous fumes. Add small amount of lead oxide (0.02g) and place a strip of fluorescein paper across the mouth of the test tube. Heat the contents to boil. A pink colour on the fluorescein strip (eosin formed) indicates bromine. Boil off bromine completely, cool and filter. Acidify the filtrate with dilute nitric acid and then add silver nitrate solution (1.0 mL). Appearance of curdy white precipitate, soluble in aqueous ammonia confirms chlorine.

$$2\,NaBr + PbO_2 + 4CH_3COOH \rightarrow Br_2 + (CH_3COO)_2Pb + 2CH_3COONa + 2H_2O$$

Note

- Ensure complete removal of iodine before testing for bromine.
- If iodine is absent, then take Lassaigne's extract (1.0 mL), acidify with glacial acetic acid, add small amount of lead dioxide and proceed as above to test for bromine.
- Lead dioxide in acetic acid forms lead tetraacetate which oxidizes hydrobromic acid to bromine but does not affect hydrochloric acid.

6. Alcoholic silver nitrate test: This test is used to distinguish aryl halides from alkyl or aryl-alkyl halides. It is based on the higher mobility of halogen attached to aliphatic carbon as compared to that attached to the ring. Alkyl or aryl-alkyl halides react with alcoholic solution of silver nitrate to give white or yellow precipitate of silver halides. Since the reaction proceeds *via* an S_N^1 mechanism, the foremost factor that will contribute to the reactivity of an alkyl halide (RX) or aryl-alkyl halide toward silver nitrate is the stability of the carbocation (R^+) that is formed. The more stable the carbocation, the faster will be the reaction with alcoholic $AgNO_3$.

$$RX \xrightarrow{Ag^+} R^{\delta+}\text{---}X^{\delta-}\text{---}Ag^{\delta+} \longrightarrow AgX\downarrow + R^+$$

$$R^+ \xrightarrow{NO_3^-} RONO_2$$

Procedure: Take the organic compound (1-2 drops or 0.05 g) in a test tube and add 2% ethanolic solution of silver nitrate (2.0 mL), dropwise with shaking. If no precipitate is formed in cold, heat the mixture on a boiling water bath for 5-10 minutes. Add 5% aqueous nitric acid (1-2 drops, do not add concentrated nitric acid). A yellow or white precipitate indicates presence of aryl-alkyl or allyl or tertiary or secondary alkyl halide.

Note

- Some organic acids precipitate insoluble silver salt which are soluble in dilute nitric acid. Therefore, persistent yellow or white precipitate after addition of dilute nitric acid confirms the presence of aryl-alkyl or alkyl halide.
- Aryl and vinyl halides will not give this test.

Safe Waste Disposal

- Transfer the test solution to a beaker and add a saturated solution of sodium chloride. Filter the precipitate of the silver halide and transfer the residue to non-hazardous inorganic waste container. Transfer the filtrate to the aqueous waste solution container.

4.3.1.6.5 *Spot tests and green methods*

Numerous green approaches are now available for detection of extra elements. These include: (a) Green preparations replacing sodium fusion test. (b) Spot and groove tile tests, which minimize the use of chemicals and generate minimum chemical waste. Three such approaches are discussed in this section.

1. Middleton's safe and green procedure*-An alternative to sodium fusion test: Metallic sodium used for fusion with organic compound, in the conventional method, is hazardous, unsafe and is a cause of great concern in a student laboratory. The idea of fusion with sodium metal is to convert the water-insoluble organically bound extra elements to water-soluble sodium salts which can be easily detected by various tests.

A better and green alternative for elemental analysis involves heating the organic compound containing extra elements with a mixture of zinc dust and sodium carbonate. The nitrogen and halogens present in the organic compound get converted into soluble zinc cyanide or zinc halide, which can be identified by following the same procedure as in Lassaigne's sodium test for nitrogen or halogen. Organic sulphur, however, forms insoluble zinc sulphide which is decomposed with hydrochloric acid and the liberated hydrogen sulphide gas is identified with lead acetate paper.

Green components of the method:

- Use of hazardous and unsafe sodium metal is avoided.
- The compounds containing both nitrogen and sulphur do not produce thiocyanates under these conditions. Hence, nitrogen and sulphur can be identified independent of each other.

*Middleton (Analyst, 1935, **60**, 154)

Drawbacks over sodium-fusion extract

- The reagents required are not obtained in pure form easily. A blank test with distilled water should always be performed side by side and the results compared to rule out any discrepancy.

Procedure: *Preparation of the Middleton's reagent*: Grind thoroughly together in a dry mortar, pure anhydrous ("Analar") sodium carbonate (2.5 g) and pure zinc dust (5.0 g). Preserve the reagent in a wide-necked stoppered bottle until required.

Take the organic compound (0.2 g or 2-3 drops, if liquid) in a small dry hard-glass test tube, add Middleton's reagent to get about 1 cm column. Close the test tube with a stopper and shake vigorously until the contents are properly mixed. Now add more reagent until a total height of about 3-4 cm is obtained (Do not mix this time!). Hold the test tube horizontally with the help of a test tube holder and heat gently at the open end, using the Bunsen burner. Heat until the mixture turns red hot at the open end. Now slowly proceed towards the closed end. Continue heating until entire mixture turns red hot. Heat for another 2-3 minutes and plunge the tube vertically into the porcelain dish containing distilled water (20.0 mL). Boil the contents gently, cool slightly and then filter. Retain the residue on the filter paper for testing sulphur and use the filtrate for testing nitrogen and halogens.

Note

- Zinc powder used for the reaction should be free from halogens, sulphur or nitrogen impurities.
- The fusion tube must be heated strongly to red hot for minimum two minutes, before plunging in china dish containing distilled water.

- If during heating, the mixture bumps out of the tube by the evolution of gas, remove the test tube from the flame and redistribute the contents by shaking vigorously. Then, reheat the test tube cautiously.

Test for nitrogen: Take the filtrate (2.0 mL) and add 10% aqueous sodium hydroxide solution (2-3 mL) and mix. Add ferrous sulphate (0.2 g) and proceed as for Lassaigne's sodium test for nitrogen. (See *section 4.3.1.6.1, page 103*)

Note

- During acidification, add dilute sulphuric acid dropwise with care as the zinc carbonate present in the solution causes vigorous evolution of carbon dioxide gas.

Test for halogens: Take the filtrate and proceed as for Lassaigne's sodium test for halogens.

Note

- Do not forget to remove cyanides, if present, before testing for halogens (see *section 4.3.1.6.4, page 107-108*).

Test for sulphur: Moisten a piece of filter paper with lead acetate solution. Take residue (0.05 g) in a boiling tube, add dilute hydrochloric acid (10.0 mL) and immediately cover the mouth of the boiling tube with the moistened paper. If sulphur is present, hydrogen sulphide gas is evolved, which gives a definite black colouration of lead sulphide on the paper.

Note

- The evolution of hydrogen sulphide can also be confirmed by its rotten egg smell.

2. Potassium or Sodium Salts of Organic Acids-An alternative to Lassaigne's sodium test: As discussed in Middleton's test, replacement of hazardous sodium metal is key to green conversion of organically bound extra elements to aqueous solution containing cyanide, halide or sulphide ions. This can also be achieved by heating the organic compound with twice the quantity of sodium or potassium salt of oxalic, succinic, benzoic or phthalic acid. The sodium or potassium cyanides, sulphides or halides, in aqueous solution, may then be identified by same procedure as in Lassaigne's sodium test.

Procedure: In a clean and dry test tube, take the given organic compound (0.2 g) and add sodium salt of carboxylic acid (0.4 g). Heat the mixture to red hot. Rotate the test tube to ensure uniform heating. Plunge the test tube vertically to the porcelain dish containing distilled water (10.0 mL). Boil the mixture for 1-2 minutes and filter. Use the filtrate to perform the test for nitrogen, sulphur or halogens as discussed in *sections 4.3.1.6.1* to *4.3.1.6.4*.

3. Spot Tests for Extra elements: Conventionally, elemental analysis and functional group identification is carried out on microscale. Even though the microscale analysis involves small amounts of chemicals, it still suffers from the following drawbacks:

- Generation of large quantities of chemical waste.
- Some of the tests involve use of hazardous organic solvents for extraction.
- Use of larger quantities of corrosive chemicals.
- LPG, a non-renewable fuel, is used for heating.

A greener and eco-friendly alternative can be found in **spot tests**. The spot tests are carried out on any one of the following: TLC plate/Merck silica plate or Whatman filter paper strip. Microscale tests may also be performed in grooved tiles. The tests are ecofriendly as they help to cut down on use and disposal of chemicals. They are energy effcient as the heating on non-renewable LPG is replaced by heating on the oven and the procedures are almost non-hazardous.

The following general instructions may be used while carrying out the spot tests:

1. Use concentrated Lassaignes's extract as well as concentrated solution of organic compounds obtained by dissolving 3-4 drops of liquid or 0.3-0.4 g of solid compound in 1-2 mL of solvent. Depending upon the solubility, the solvent should be chosen in the order: water, alcohol, acetone and finally ethyl acetate. Do not use acetone for 2,4-DNP test, ethyl acetate for hydroxamic acid test and alcohol for ceric ammonium nitrate (CAN) test or Baeyer's unsaturation test.

2. Concentrated aqueous solutions of water-soluble compounds can be prepared except for hydroxamic acid test.

3. Take a 1" × 6" Whatman paper strip to perform the tests except for the test involving concentrated sulphuric acid, as the paper gets charred.

4. Feed the sample and the reagents using a narrow capillary tube.

5. Spot from one end of the Whatman paper so that judicious use of Whatman paper is possible.

6. Spotting of the reagents should be done one over the other sequentially as given in the procedure.

7. Use separate capillaries for each reagent. Do not mix with one another.

8. The reagent bottles should always be covered to prevent loss of water due to evaporation.

9. Heating should be done in a pre-heated electric oven maintained at ~120°C.

10. Replace the reagents, if discolored.

Spot Test For Nitrogen (Lassaigne's Sodium Test): Spot the following solutions one over the other, in the order given below, on a thin strip of Whatman filter paper:

- Concentrated Lassaigne's extract (spotted twice)
- A saturated aqueous ferrous sulphate solution (spotted twice).
- Dilute sulfuric acid

Note

- Boil the ferrous sulphate solution for 2-3 minutes before spotting. This ensures both ferrous and ferric ions are present in the solution.
- Appearance of a deep blue coloured spot confirms the presence of nitrogen in the unknown compound.

Spot Test for Sulphur (Nitroprusside Test): Spot the following solutions sequentially on a thin strip of Whatman filter paper:

- Concentrated Lassaigne's extract (spotted twice)
- Freshly prepared sodium nitroprusside solution (Use concentrated solution)

Appearance of a deep purple colored spot confirms the presence of sulphur in the unknown compound.

Spot Test for Nitrogen & Sulphur (Lassaigne's Sodium Test): Spot the following solutions, sequentially, on a thin strip of Whatman filter paper:

- Concentrated Lassaigne's extract (spotted twice)
- Dilute HCl
- Concentrated $FeCl_3$ solution

Appearance of blood red colored spot confirms the presence of nitrogen and sulphur in the unknown compound.

4.3.1.7 Heating with Soda Lime

This test can supplement the knowledge gained from solubility test and provide additional information about the nature of functional group that could be present in the given organic compound. The nature of gases evolved on heating with sodalime or liquid condensate obtained serves as the basis for additional information, see Table 4.4.

Procedure: Grind powdered organic substances (0.2 g) with powdered soda lime (1.0 g), in a clean mortar. Place the solid mixture in a pyrex-glass test tube (make sure that the test tube is not more than half-full), close the mouth of the tube with the help of a cork and delivery tube and then slant the test tube as shown so that the condensate does not run back into the hot part of the test tube (see Fig. 4.2). Now heat the tube first gently and then vigorously. Any non-condensable gases such as hydrogen or methane is best detected by collecting a sample of the gas in test tube A whereas any condensable liquid such as phenol, benzene or aniline should be collected in small amount of water as shown in test tube B.

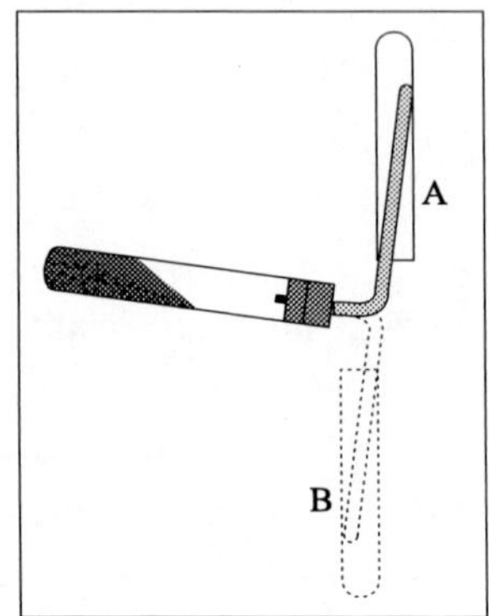

Fig. 4.2   Heating of Compound with Soda Lime

Table 4.4   Interaction of Unknown Compound with Sodalime

S. No.	Observation	Inference
1.	Ammonia evolved on heating	Amides
2.	Benzene like smell	Aromatic carboxylic acid
3.	Aniline, fish like odour (perform Azodye test)	Aromatic amino carboxylic or sulphonic acids
4.	Phenol (coloration with $FeCl_3$)	Phenolic acids
5.	Fishy smell of amines on prolonged heating	Anilides
6.	Smell of burning sugar	Carbohydrate or polyhydroxy compounds

4.3.1.8   *Unsaturation Test*

Organic compounds containing either C = C or C ≡ C or aromatic compounds having side chains with C = C or C ≡ C, decolourise a solution of bromine in a suitable solvent (water/acetic acid or CCl_4) and Baeyer's reagent (alkaline potassium permanganate). Aromatic compounds like benzene, toluene, xylenes and naphthalene in spite of having high degree of unsaturation do not respond to these tests. These reactions are based on the fact that the unsaturated compounds can easily undergo addition reaction while the aromatic compounds due to high degree of stability associated with the aromatic character do not show addition reaction under normal conditions. The presence of unsaturation in an organic compound can be established by doing the following tests:

1.   Baeyer's reagent test: Cold dilute and alkaline solution of potassium permanganate oxidizes the unsaturated hydrocarbons, aromatic compounds

$$3 \ \text{C=C} \ + 2KMnO_4 + 4H_2O \longrightarrow 3 \ \underset{\text{OH OH}}{\text{C--C}} \ + 2\,MnO_2 + 2KOH$$

Alkene
Diol

$$R\text{---}\!\!\equiv\!\!\text{---}R + 2KMnO_4 \longrightarrow 2\,RCOOK + 2\,MnO_2$$

Alkyne
Potassium carboxylate

Phenol $\xrightarrow{[O]}$ (hydroquinone) $\xrightarrow{[O]}$ Quinone

Phenol **Quinone**

$$RCHO \xrightarrow[\text{(ii) H}^+]{\text{(i) KMnO}_4/\text{OH}^-} RCOOH$$

Aldehyde **Carboxylic acid**

having multiple bonds in side chain, aldehydes, some alcohols, carbohydrates, formic acid and its ester, aromatic amines and phenols. The reaction is accompanied by disappearance of the purple colouration and formation of a brown precipitate of MnO_2.

Procedure: Dissolve solid (0.1 g) or liquid (0.2 mL) in water or acetone (do not use ethanol) in a clean test tube, add 1% sodium carbonate solution (1-2 drops), followed by addition of dilute solution of potassium permanganate (2-3 drops, 1-2%). Immediate discharge of pink colour indicates presence of unsaturation or easily oxidisable group in the unknown organic compound.

Note

- Oxalates reduce acidified and alkaline potassium permanganate only on warming
- Formates reduce both acidic and alkaline potassium manganate in cold.

2. Decolourisation with bromine solution: Compounds containing C = C or C ≡ C or aromatic compounds with multiple bonds in the side chain, aromatic amines and phenols decolourise bromine solution. This test can be used to distinguish between compounds containing multiple bonds (alkenes, alkynes) from those containing easily oxidisable group. For example, aldehydes, phenols and aromatic amines. Compounds having carbon-carbon multiple bonds (double bond and triple bond) as well as arenes having multiple bonds in the side chain undergo addition reaction with bromine solution forming vicinal dihalides or tetrahalides, with disappearance of the brown colour of bromine solution.

$$\text{Alkene} + 2\,Br_2 \longrightarrow \underset{Br \quad Br}{\text{(Brown colour discharge)}}$$

Phenols and aromatic amines, due to the presence of strong activating groups, undergo electrophilic substitution reaction with bromine solution to precipitate white coloured polybromo derivative, with evolution of HBr gas and disappearance of the brown colour of bromine solution.

Procedure: Take solid (0.1 g) or liquid (0.2 mL) in a clean test tube and add CCl_4 or water or acetic acid (2.0 mL, depending upon the solubility of the compound). Add dropwise, with shaking, a solution of bromine (5%) in the same solvent as taken for the unknown compound. Discharge of brown colouration indicates unsaturation whereas discharge of brown colour along with appearance of white precipitate, with or without evolution of HBr, indicates phenol or aromatic amine (please correlate with extra element test and solubility).

$$OH + 3Br_2 \xrightarrow{H_2O} \text{2,4,6-Tribromophenol} + 3HBr$$

2,4,6-Tribromophenol
(White ppt.)

$$NH_2 + 3Br_2 \xrightarrow{\text{Acetic acid}} \text{2,4,6-Tribromoaniline} + 3HBr$$

2,4,6-Tribromoaniline
(White ppt.)

Note

- This test is to be performed only if Baeyer's reagent test is positive.
- A positive Baeyer's test and negative bromine test confirms the presence of easily oxidizable group.
- Bromine is hazardous, highly corrosive and has pungent smell. It should be handled with care. Do not perform this test without hand gloves on.
- Rate of bromination is enhanced in polar solvents like water and acetic acid.

- Presence of electron withdrawing or bulky group next to multiple bonds decrease the reaction rate, e.g., cinnamic acid reacts slowly whereas 1,1,2,2-tetraphenylethylene does not show any addition reaction with bromine in CCl_4.

- Evolution of HBr, in case of phenols and aromatic amines, will be observed only when CCl_4 is used as solvent. Evolution of HBr can be confirmed by bringing blue litmus paper near the mouth of the test tube and noting whether it turns red due to evolution of an acidic gas. Formation of anilinium bromide with liberated HBr may prevent evolution of HBr in aniline and substituted anilines. Since HBr is soluble in water, no HBr evolution is seen in cases where water is used as solvent.

Safe Waste Disposal

- Place the test solution in the hazardous waste container.

3. Spot test for unsaturation (Baeyer's reagent test): Take concentrated solution of the compound in water or acetone and spot the following one over the other, sequentially:

- Concentrated solution of the compound

- Baeyer's reagent

Disappearance of pink colour indicates presence of unsaturation/easily oxidisable group.

Note

- Pink colour should be discharged immediately.

- Always perform a blank test to rule out any discrepancy.

4.3.1.9 *Melting Point and Boiling Point*

Determination of the correct melting or boiling point for an unknown compound (within +/- 2-3°C) is of critical importance, as it not only tells you about the purity of the sample but it is also used as a means of identification of unknown organic compounds based on classified tables of organic compound listed in all organic practical books. A classified table of organic compounds in Appendix I, in this book, also provides this information. In general, a compound having sharp melting point (melts within 1°C range) or sharp boiling point is considered to be pure (***Criterion for Purity**, Chapter 4*, Volume 1, for more details). In case you have accurately determined the melting or boiling point of the compound containing a particular functional group, then you have made your task easier as your compound now lies in the list of three to four compounds having melting/boiling point in that range.

Post Lab Questions

1. List some physical properties to distinguish between an inorganic compound and an organic compound.

2. What is the role of preliminary tests in qualitative analysis of an unknown organic compound. Explain taking examples of the following:
   (a) Solubility test
   (b) Flame test
   (c) Baeyer's reagent test

3. Certain organic compounds burn with sooty flame whereas certain other give off non-sooty flame. Explain.

4. Can you think of some aliphatic compounds that burn with sooty flame? Name two such compounds.

5. "Like dissolves like". Explain this phrase in context of the solubility of organic compounds. Give reasons for your answer.

6. If a compound is soluble in cold water, what other solubility tests will you perform and why?

7. Name the categories of organic compounds which will dissolve in the following and give reasons for your choice:
   (a) Dilute NaOH
   (b) Dilute $NaHCO_3$
   (c) Dilute HCl

8. Why is it important to test for extra elements before performing the solubility test in concentrated sulphuric acid?

9. Define "extra elements" and give examples.

10. What is Lassaigne's extract and why do you need to prepare this before testing for extra elements?

11. Sodium is a very reactive metal. List the precautions one should take while preparing the Lassaigne's extract. Can you suggest any methods to replace sodium, while extracting extra elements in aqueous solution?

12. Can you use potassium in place of sodium to prepare Lassaigne's extract? Give reasons.

13. A student left sodium metal uncovered, open in air. Do you think it was a right thing to do. If not, why?

14. It is always advised to take a fresh, shiny piece of sodium for preparing Lassaigne's extract. Why do you think this is important?

15. Lassaigne's extract should never be coloured. Give reasons and explain what one should do in that case?

16. Why one should ensure that Lassaigne's extract is alkaline before testing for nitrogen?

17. Why do you need to acidify Lassaigne's extract while performing the following tests:

    (a) Prussian blue test for N

    (b) Lead acetate test for S

    (c) Silver nitrate test for halogens

    (d) Layer test for halogens

18. Can you use dilute HCl or dilute HNO_3 while doing the sodium fusion test for nitrogen? If not, why?

19. What is Prussian blue and how does it differ from Turbull's blue? Explain the formation of Prussian blue complex while doing sodium fusion test for nitrogen.

20. Sometimes while performing sodium fusion test for nitrogen, green colour is obtained instead of Prussian blue. What does that indicate?

21. Which of the following will give positive sodium fusion test for nitrogen and why?

    (a) Hydrazine/phenyl hydrazine

    (b) Hydroxylamine/sodium cyanide

    (c) Benzene diazonium chloride/N, N-diphenylhydrazine

22. A student while performing sodium fusion test for nitrogen, takes Lassaigne's extract, adds a pinch of ferrous sulphate, shakes and acidifies with dilute sulphuric acid. What could be the possible colouration/precipitate? Pick out the best observation and reason from the following:

    (a) He got Prussian blue.

    (b) He did not get Prussian blue because Lassaigne's extract was not alkaline.

    (c) He did not get Prussian blue because he did not heat, after adding ferrous sulphate.

    (d) He did not get Prussian blue because he did not heat after adding ferrous sulphate, so there were no ferric ions in solutions to give Prussian blue colouration.

23. Why is it important to remove N and S before testing for halogen in the Lassaigne's extract?

24. Baeyer's test is not a confirmatory test for the presence of unsaturation. Explain.

25. Which of the following compounds will decolorize bromine solution and why? Give the major product in each case. Also explain why reaction is slower in option "c".

    (a) Phenol   (b) Aniline

    (c) Cinnamic acid   (d) Styrene

    (e) Toluene   (f) 1,3-butadiene

    (g) Benzaldehyde   (h) Naphthalene

26. Why ethanol should not be used as a solvent while performing the Baeyer's reagent test?

27. Can you use dilute sulphuric acid for acidification in lead acetate test for sulphur? If not, why?

28. Which of the following will give Beilstein test and why?

    (a) Chlorobenzene

    (c) Urea

    (b) Benzyl chloride

    (d) Benzonitrile

29. How will you test for nitrogen in hydrazine?

30. What is Liebig's test?

31. A student performs combination test for N and S on Lassaigne's extract of thiourea. He did not get blood red colouration. Where did he go wrong?

32. What is the need to do the N and S combination test?

33. What is Layer's test and why is it performed?

34. What is the role of concentrated nitric acid in layer test. Can we use potassium permanganate in place of nitric acid. If yes, what changes are required in the procedure?

35. Give one preliminary test to distinguish between the following:

    (a) Benzaldehyde and acetophenone

    (b) Benzaldehyde and acetaldehyde

    (c) Chlorobenzene and benzyl chloride

36. Stilbene decolourises bromine water very slowly as compared to styrene. Why?

37. How does bromination of aniline differ from that of styrene?

38. Why is sodium kept under kerosene in the lab and not water/alcohol?

39. Explain why dilute HNO_3 is added during silver nitrate test for halogens. Can you use dilute HCl or dilute H_2SO_4?

4.3.2 Test for Functional Groups and Derivative Preparation

A functional group refers to an atom or group of atoms which characterize the physical and chemical properties of an organic compound. Characterization of the functional group is a crucial step in the identification of an unknown organic compound. If you have performed all the tests listed under *section 4.3.1*, then you have a fair idea of the physical properties and presence/absence of extra element in the given unknown organic compound and also the class of the organic compound. The solubility tests along with the information gained from elemental analysis will help you categorize your compound into one of the three broad categories—acidic, basic and neutral. Now it is time for you to carry out a deeper analysis through a few classification tests to characterize the functional group(s) that you suspect could be present in the unknown compound. If you judiciously perform the classification tests and analyze the IR and NMR data, you will be able to characterize a significant fraction of organic compounds. The tests have been listed under the specific functional group category to which they apply. For the convenience of the readers, we have included some of the simplest and most reliable of the numerous classification tests that have been developed till date. It is suggested that the students should go through the data obtained thus far and plan prudently as to what functional group tests to perform. You must not run all these characterizing chemical tests. First test in each functional group category is the *group test* which is followed by *special/ confirmatory tests*. You should perform special tests only if you have a positive group test. You should be smart enough to choose and reject a particular test. For example, if your solubility data suggests that your compound is neutral, you should not waste time doing test for carboxylic or phenolic groups.

In case your compound contains more than one functional group then the classification and characterization should be done on the basis of a functional group which is easily manipulated and detected. For example, *p*-hydroxybenzaldehyde, *p*-methoxybenzaldehyde and *p*-chlorobenzaldehyde will all be classified as aromatic aldehydes. However, it is important that the subsidiary functional group is also identified before the derivative preparation is attempted. Gauge your suppositions with specific tests before you arrive at a conclusion. You should first try control test on known compounds such as benzaldehyde and phenol, in the above case, so that you can observe directly both positive and negative results for the test. Likewise, if you are not familiar with the smell of burning sugar, it is worthwhile to take glucose or fructose and see how they smell on burning. The frustration of ambiguous results can be minimized by doing control tests on reference compounds.

Note

- Perform the classification tests systematically and in the order listed in the *sections 4.3.2.1* to *4.3.2.6*.

- Some compounds do not give a positive group test for a functional group present in them. In such cases, you must use the information gained from *section 4.3.1* along with the spectral data to arrive at a logical conclusion. For example, 1-/2-naphthol does not give any characteristic coloration in aqueous ferric chloride solution but will be soluble in 5% aqueous NaOH and will not give effervescence with 5% aqueous $NaHCO_3$ solution.

- Sometimes compounds appear to give test for the functional groups not present in them. For example, aniline gives a yellow precipitate with 2,4-dinitrophenylhydrazine (2,4-DNP), sometimes, not due to 2,4-dinitrophenylhydrazones but due to salt formation with the HCl present in the 2,4-DNP reagent. If, the precipitate is decanted and washed, it dissolves in excess of water. It is also noticed that the colour of the precipitate is white and not yellow. It is perceived as yellow due to yellow-orange colour of the 2,4-DNP reagent. You should therefore be judicious while concluding the results for any chemical tests.

4.3.2.1 Compounds Containing Carbon, Hydrogen and Oxygen (Acidic Compounds)

These are compounds containing one or more acidic groups. These include (1) carboxylic acids (2) sulphonic acids (3) phenols (4) thiols, etc. However, for the undergraduate students, only carboxylic acids and phenols are discussed in this book. Both compounds are soluble in aqueous sodium hydroxide and the distinction between the two can be made on the basis of their behaviour with aqueous sodium bicarbonate solution. Carboxylic acids, being the stronger acid, dissolve in aqueous sodium bicarbonate solution with evolution of carbon dioxide whereas most phenols are generally not soluble and do not evolve carbon dioxide, being weaker acids. (See the discussion on phenols, *section 4.3.2.1.2, page 135,* for details).

4.3.2.1.1 Carboxylic acids

These are compounds having carboxyl group (–COOH). They are colourless crystalline solids except formic acid, acetic acid and lactic acid, which are liquid at room temperature. Small chain aliphatic acids are soluble in cold water. The aromatic acids like benzoic acid and toluic acid, etc., are sparingly soluble in cold water, but dissolve readily in hot water. Phthalic acid, having two carboxyl groups, is more soluble in cold water than the other aromatic acids. Carboxylic acids are commercially used in production of polymers, pharmaceuticals, solvents and food additives. For example, adipic acid is used for nylon synthesis, citric acid is a food flavouring agent and a preservative. Fatty acid salts with alkali metals are used as soaps as they have detergent properties.

The most significant property of carboxylic acids is their acidity. They ionize in aqueous solution to liberate protons and hence turn blue litmus red. The extent of acidity depends on the ease with which they can ionize in solution (pK_a values). Lower the pK_a value, higher will be the acidic character. See Table 4.5 for a list of commonly found carboxylic acids in the undergraduate chemistry lab, along with their pK_a values. Table 4.6 lists the common classification tests for carboxylic acids.

Table 4.5 Commonly found Carboxylic Acids in Chemistry Lab

S.No.	Name of the carboxylic acid	PK_a value (at 25°C)*	GHS Symbols**
1.	Acetic acid	4.76	
2.	Benzoic acid	4.20	
3.	o-Toluic acid	3.91	
4.	Oxalic acid	1.46 and 4.40	
5.	Phthalic acid	2.76 and 4.92	
6.	Succinic acid	4.21 and 5.7	
7.	Trans-Cinnamic acid	4.44	
8.	p-Nitrobenzoic acid	3.44	

Source: https://pubchem.ncbi.nlm.nih.gov; **Please refer Volume 1, Chapter 1, for details on GHS safety symbols.

Table 4.6 Classification Tests for Carboxylic Acids

S.No	Classification Test	Test category	Observation	Inference
1.	Solubility in 10% aqueous NaOH	Preliminary Test	Soluble. Precipitates on acidification with concentrated HCl	Acidic Compound (Phenol/Carboxylic acid)
2.	Litmus paper test*	Preliminary Test	Blue litmus turns red	Acidic Compound (Carboxylic acid or some substituted phenols)
3.	Saturated aqueous NaHCO$_3$ Test	Group test	Soluble with brisk effervescence	Carboxylic acid
4.	Esterification Test	Special test	Fruity smell	Carboxylic acid
5.	Fluorescein Test	Special test	Green Fluorescence	Dicarboxylic acid capable of forming stable cyclic anhydride

*Should be done for water soluble compounds only.

Classification tests for carboxylic acids

1. Solubility in 10% aqueous NaOH: See *section 4.3.1.5*, page 95, solubility tests, for details.

Note

- Phenols are also soluble in 10% aqueous NaOH. Sometimes coloured solutions are obtained on dissolution, due to formation of phenolate ion (better auxochrome). For example, nitrophenols.

Safe Waste Disposal

- Carefully neutralize the test solution with 5% aqueous HCl, with stirring, until a pH ~7 is attained (use pH paper). Transfer the contents to aqueous waste container.

2. Litmus paper test: Carboxylic acids liberate protons in aqueous solution and therefore turn blue litmus paper/solution to red.

Procedure: Take aqueous solution of unknown compound (1.0 mL, some acids may require heating to dissolve), insert a glass rod in the solution and introduce a drop of aqueous solution on a small strip of blue litmus paper. Appearance of red colouration confirms that the compound is acidic. Some phenols also turn blue litmus to red due to their acidic character. For example, 2,4-dinitro and 2,4,6-trinitrophenol.

Note

- Litmus is a weakly acidic organic dye which is red below pH 4.5 and blue above pH 8.3. Neutral litmus paper is purple.

3. Sodium bicarbonate test: Carboxylic acids react with aqueous sodium bicarbonate or sodium carbonate solution to form the carboxylate anion and carbon dioxide gas. The liberation of carbon dioxide is seen in the form of brisk effervescence.

$$\underset{\text{Carboxylic acid}}{\text{RCOOH}} + \text{NaHCO}_3 \longrightarrow \underset{\text{Sodium carboxylate}}{\text{RCOONa}} + \text{H}_2\text{O} + \text{CO}_2$$

Procedure: Prepare a saturated solution of sodium bicarbonate in water (1.0 mL) and transfer it to a watch glass. Add the given organic compound (0.1 g of solid or 0.2 mL liquid). Compound dissolves with generation of brisk effervescence, indicating the presence of carboxylic acid.

Note

- Sometimes the release of carbon dioxide is slow due to poor solubility of acid in water. In such cases, warm water may be taken for preparing saturated sodium bicarbonate solution (do not use boiling water). Else, add solution of compound in methanol to saturated sodium bicarbonate solution, when brisk effervescence is evolved.

- Water soluble phenols also dissolve in aqueous $NaHCO_3$ solution, however no evolution of CO_2 takes place, *i.e.*, without the formation of a sodium derivative. This reaction can therefore be used to distinguish between carboxylic acids and most phenols. (See *section 4.3.1.5, page 93, solubility tests*).

- Perform blank test, if alcohol is used as solvent. This rules out the presence of any acidic impurity in alcohol.

- Certain substituted phenols, mainly nitro phenols, are adequately acidic to liberate CO_2 from aqueous $NaHCO_3$. However, all nitro phenols, give red or yellow solutions with aqueous sodium bicarbonate due to precipitation of sodium salt.

Safe Waste Disposal

- Transfer the test solution to aqueous waste container after neutralizing to pH ~7 with 5% aqueous HCl.

4. Ester formation: Carboxylic acids react with alcohols in the presence of mineral acids, as catalyst, to form pleasant smelling esters. This reaction forms the basis of their identification in the lab.

$$\underset{\text{Carboxylic acid}}{RCOOH} + \underset{\text{Ethanol}}{C_2H_5OH} \xrightarrow[\text{heat}]{\text{Conc. } H_2SO_4} \underset{\text{Ester}}{RCOOC_2H_5} + H_2O$$

Procedure: Take the given organic compound (0.05 g), alcohol (4-5 drops) and concentrated sulphuric acid (2-3 drops) in a dry test tube, put a glass rod and heat the contents on a boiling water bath for 2-3 minutes. Pour the contents carefully, with stirring, into a beaker containing water (~20-25 mL). Neutralize with 10% aqueous sodium hydroxide solution. Note the pleasant smell due to formation of ethyl ester.

Safe Waste Disposal

- Carefully neutralize the solution with 5% aqueous HCl, with stirring, until a pH ~7 is attained (use pH paper) and transfer to aqueous waste container.

5. Fluorescein test: This test can be used for distinction between monocarboxylic acids like benzoic acids, toluic acid and some dicarboxylic acids capable of forming stable cyclic anhydrides. For example, succinic acid, and phthalic acid. According to **Blanc's rule,** 1,4 and 1,5-dicarboxylic acids, when heated, form stable cyclic anhydrides.

When resorcinol is heated in dry conditions with 1,4 or 1,5-dicarboxylic or their cyclic anhydride, in the presence of concentrated sulphuric acid, as catalyst, it forms **fluorescein**, a fluorescent dye which gives green fluorescence in alkaline medium.

COOH
COOH

Conc. H_2SO_4
$-H_2O$

O
O

Phthalic acid

Phthalic anhydride

Phthalic anhydride + 2 Resorcinol

H_2SO_4
$-2H_2O$

Fluorescein

NaOH
$-H_2O$

Green Fluorescence

Procedure: Take a clean and dry test tube and add the given organic compound (0.05 g), resorcinol (0.05 g) and concentrated sulphuric acid (3-4 drops). Insert the glass rod and heat the contents carefully on a small Bunsen burner flame, until the contents fuse/melt. Transfer the contents carefully to a beaker containing dilute sodium hydroxide solution (30-40 mL, 3-5%). Appearance of yellow-orange solution with green fluorescence confirms the presence of a dicarboxylic acid capable of forming stable cyclic anhydride.

Note

- Oxalic acid and adipic acid will give a –ve fluorescein test (Why!)
- Do not overheat the mixture as the compound chars and does not give the desired results.
- The solution after transferring to sodium hydroxide solution should be alkaline (check pH)

Safe Waste Disposal

- Neutralize the test solution with 5% aqueous HCl until neutral to pH paper and discard in the aqueous waste container.

6. Special test for oxalic acid

Oxalic acid decolorizes acidified potassium permanganate due to reduction of Mn (+7) to Mn (+2). This reaction can be used to confirm the presence of oxalic acid.

Procedure: In a clean and dry test tube, take aqueous solution of the given compound (1.0 mL), add dilute sulphuric acid (2N, 2-3 drops) and warm the solution on wire gauze to about 60-80°C. Add dilute solution of potassium permanganate (2-3 drops). Note instant decolorization due to reduction of Mn (+7) to Mn(+2).

$$2KMnO_4 + 3H_2SO_4 + 5H_2C_2O_4 \longrightarrow K_2SO_4 + 2MnSO_4 + 10CO_2 + 8H_2O$$

Note

- This test is performed only if sodium bicarbonate test is positive and fluorescein test is negative.
- Do not boil the solution while performing the test.

7. Spot tests for Carboxylic acids

(a) Sodium bicarbonate test

- Prepare strong concentrated aqueous solutions of your compound and sodium bicarbonate.
- Take a capillary and dip the capillary into concentrated solution of your compound.
- Some solution will rise up the capillary due to capillary action.
- Remove the capillary and dip it next into concentrated aqueous solution of sodium bicarbonate.

Brisk effervescence in the capillary confirms the presence of carboxylic acid.

(b) Fluorescein Test

Prepare strong concentrated aqueous solutions of your compound, sodium hydroxide, resorcinol and spot them one over the other, two times each, on a TLC/Merck plate in the sequence mentioned below:

- First spot the concentrated aqueous solution of your compound.
- Next, spot the concentrated aqueous solution of resorcinol.
- Next introduce 2 drops of concentrated sulphuric acid.
- Heat the plate in oven (maintained at ~120°C) for 40-50 seconds, till the spots dry up completely.
- Scratch the spotted portion of the gel and introduce it into a test tube containing sodium hydroxide solution (5-7 drops).

A fluorescent green colour in the test tube confirms presence of 1,4 or 1,5-dicarboxylic acid.

8. Spectroscopic identification of carboxylic acids

Table 4.7   Spectral Data of Carboxylic Acids

S. No.	Type of spectrum	Peak value	Peak attribution	Remarks
1	IR	1690-1760 cm^{-1}	C=O Str.	Strong, sharp
		2500-3300 cm^{-1}	O-H str. for COOH	Medium, broad
2	NMR	10-13 δ*	COO**H**	Replaceable with D$_2$O
		2-3 δ	α-H to COOH	-

*Position of acidic hydrogen changes with concentration of acid taken.

Derivatives of Carboxylic acids: Carboxylic acids can be converted into a variety of derivatives like anhydrides, amides, *S*-benzylisothiouronium salts, anilides, etc. Some of the common ones are discussed below.

1. S-Benzylisothiouronium salt: *S*-Benzylisothiouronium salts (SBT) are extensively employed to identify carboxylic acid. *S*-Benzylisothiouronium salt of benzoic acid is synthesized by reaction of sodium salt of benzoic acid with *S*-benzylisothiouronium chloride in faintly alkaline medium. These compounds are crystalline and have sharp melting point, when pure.

S-Benzylisothiouronium chloride   Sodium carboxylate   S-Benzylisothiouronium carboxylate

Procedure: Take a clean and dry boiling tube. Add carboxylic acid (0.25 g) and aqueous sodium bicarbonate solution (added dropwise) until effervescence of carbon dioxide cease. Now, transfer the contents to a solution of SBT (0.5 g) in water (5.0 mL). Cool the solution in ice bath. Filter and dry the precipitated solid. Recyrstallize a fraction of the crude sample using hot water or aqueous ethanol. Filter the crystallized solid, dry it on a porous clay plate and record its melting point.

Alternatively, in a clean and dry boiling tube, dissolve carboxylic acid (0.25 g) in minimum amount of hot water (2-3 mL). Add methyl orange (2-3 drops) followed by the addition of 5% aqueous sodium hydroxide until the colour changes to yellow-orange. Immediately add dilute HCl (0.1N) dropwise until the yellow/orange colour of the solution begins to fade. Pour this solution dropwise to a solution of S-benzylisothiouronium chloride (0.5 g in 5.0 mL water). Cool the solution in ice bath for 10-15 minutes. Filter the precipitated solid. Dry the solid and recrystallize a part of crude sample using water or aqueous ethanol. Filter the crystallized solid, dry on a porous clay plate and record its melting point. (For details, see SBT preparation in our book Volume 1, *section 6.16.1*, acid derivatives).

Note

- Any undissolved acid will dissolve when the solution is made alkaline. This happens due to formation of carboxylate salts.

- Certain polybasic acids and amino acids do not from satisfactory carboxylates.

- SBT derivatives are usually pure and do not require recrystallization.

- Avoid adding excess of acid/alkali, as SBT regent is known to decompose in strongly acidic or alkaline medium (For details, see *section 6.16.1*, Volume 1).

Safe Waste Disposal

- Transfer all the solutions to aqueous waste container after neutralizing to pH 7.

2. Amide: Carboxylic acids are known to form amides in two steps. The first step involves conversion of carboxylic acid to acid halide by treatment with phosphorus pentachloride or thionyl chloride, followed by reaction with concentrated ammonia. The two-step procedure is essential as the direct reaction of carboxylic acid with ammonia proceeds very slowly.

$$RCOOH \xrightarrow[\substack{-POCl_3 \\ -HCl}]{PCl_5} RCOCl \xrightarrow[HCl]{NH_3} RCONH_2$$

Carboxylic acid   **Amide**

For detailed theory, see *section 6.16.4*, Volume 1.

Procedure: Place the given carboxylic acid (1.0 g) in a dry china dish. Add phosphorus pentachloride (4.0 g) to it and stir with the help of glass rod until it forms a clear solution or a paste (warm the mixture on steam water bath, in fume cupboard, if the paste does not liquify). Cool the china dish in ice-water and treat the acid chloride (1.0 mL) with concentrated liquid ammonia (20.0 mL) dropwise with stirring (use a fume cupboard for addition). If no solid separates in cold, warm the contents on hot water bath for 10-15 minutes and then cool. Transfer the reaction mixture, with stirring, to ice-water (~50 mL). Filter the precipitated solid. Wash the precipitate with excess of dilute sodium bicarbonate until all effervescence cease (this removes any unreacted acid). Wash with cold water, dry the solid and recrystallize a small part of crude sample with hot water or aqueous ethanol. Filter the crystallized solid, dry on a porous clay plate and record its melting point.

Note

- Phosphorus pentachloride should be fuming, when opened, otherwise use fresh bottle.
- Phosphorus pentachloride is highly hygroscopic, do not take too much time in weighing. Refer MSDS for handling instructions (https://pubchem.ncbi.nlm.nih.gov).
- Carry out the reaction outside the lab or in fume hood as PCl_5 fumes lead to acute lung injury, while ammonia and acid chloride are irritating to eyes.
- Avoid inhaling the fumes of HCl released during the reaction.
- Amides of water-soluble carboxylic acids are also likely to be soluble in water, prepare the SBT derivatives in such cases.

Safe Waste Disposal

- Dilute the filtrate with water and add 5% aqueous hydrochloric acid until pH ~7. Flush down the drain.

3. Anilide: Carboxylic acids are known to form anilides *via* reaction of acid halide (obtained by reaction of acid with phosphorus pentachloride or thionyl chloride) with aniline.

$$RCOOH \xrightarrow[\substack{- POCl_3 \\ - HCl}]{PCl_5} RCOCl \xrightarrow[-HCl]{PhNH_2} RCONHPh$$

Carboxylic   **Anilide**
acid

Procedure: Place carboxylic acid (1.0 g) in a clean and dry china dish. Add phosphorous pentachloride (4.0 g) and stir with the help of a glass rod until a clear solution is obtained. Warm the reaction mixture on steam water bath, if required. Take aniline (0.5 mL) in clean and dry round bottomed flask and then add 2-3 drops of the prepared acid chloride (~1.0 mL) and shake the contents vigorously. When the contents come back to room temperature, add the next 3-4 drops of acid chloride and repeat the above procedure till all the acid chloride has been added. Put the air condenser and reflux on the water bath for ~10 minutes. Cool and pour the reaction mixture over crushed ice (~50.0 g), in a beaker. Follow (a) or (b) as applicable.

(a) *In case a solid is obtained*: Filter the precipitated solid. Wash the precipitate first with excess of dilute sodium bicarbonate solution until all effervescence cease and then with cold water.

(b) *If no solid is obtained*: Decant the aqueous layer and add dilute hydrochloric acid to the organic liquid and shake. This will dissolve any unreacted aniline. Again decant the aqueous layer. Next treat the residue with saturated solution of sodium bicarbonate to dissolve any unreacted acid. Finally wash the residue with water. Residue which remains is, anilide. Filter, and dry the residue.

Recrystallize a part of crude sample (a) or (b) using aqueous ethanol. Filter the crystallized solid, dry on a porous clay plate and record its melting point.

Safe Waste Disposal

- Neutralize the aqueous alkaline solution with 5% aqueous HCl until pH ~7 and transfer to aqueous waste container. Add dropwise, 5% aqueous NaOH to the acidic solution until pH~7. Extract with ether (5-10 mL). Discard the ether layer to the flammable organic waste container. Run the aqueous layer down the sink.

4. Anhydride: Cyclic anhydrides are formed by sublimation process when dicarboxylic acids having formula, $COOH\text{-}(CH_2)_n\text{-}COOH$, where n=2/3, are heated under anhydrous conditions. For example, phthalic acid on heating forms phthalic anhydride.

Phthalic acid $\xrightarrow[-H_2O]{\text{Heat}}$ Phthalic anhydride

Procedure: Place the given carboxylic acid (1.0 g) in a clean and dry china dish covered with a perforated filter paper and place the china dish over a sand bath. Mount an inverted funnel over the perforated filter paper. Loosley plug the stem of the funnel with cotton. Heat the china dish gently on a sand bath. The colourless crystals of anhydride begin to form on the inner walls of the funnel. Continue heating until entire solid has sublimed. Allow the contents to cool, collect the crystals from the inner walls of the funnel and from top of the filter paper. Determine the melting point of pure crystalline derivative.

Note

- Only dicarboxylic acids following Blanc's rule form stable cyclic anhydride. For example, succinic acid, phthalic acid, glutaric acid, etc., form cyclic anhydride but oxalic acid and adipic acid do not.

5. Acetylation of salicylic acid (Special derivative)

Salicylic acid can be acetylated with acetic anhydride in the presence of traces of concentrated sulphuric acid, as catalyst, to form aspirin.

Salicylic acid   + $(CH_3CO)_2O$ $\xrightarrow{\text{conc. } H_2SO_4}$ Acetyl salicylic acid or aspirin $+ CH_3COOH$

Acetic anhydride

For detailed theory, see Volume 1, *section 6.7.6.*

Procedure: Take a clean and dry boiling tube. Add salicylic acid (0.5 g), acetic anhydride (1.5 mL) and concentrated sulphuric acid (AR,1-2 drops). Heat the contents at 50-60°C, in a preheated water bath, for 5-10 minutes. Allow the contents to cool to room temperature. Transfer the contents to ice-water (50.0 mL). Filter the solid, wash well with cold water, dry and recrystallize from aqueous ethanol. Report the melting point of recrystallized product.

Post Lab Questions

1. Carboxylic acids are more acidic than phenols. Discuss.
2. Carboxylic acids do not give the tests characteristics of carbonyl group in spite of presence of C=O group in them. Explain.
3. Arrange the following in the increasing order of acidic character giving reasons:
    (a) Benzoic acid, p-nitrobenzoic acid, o-nitrobenzoic acid
    (b) Acetic acid, monochloroacetic acid, dichloroacetic acid
    (c) Oxalic acid and malonic acid
4. Which of the following give effervescence with sodium bicarbonate and why?
    (a) Carbolic acid
    (b) Acetic acid
    (c) p-Nitrophenol
    (d) 2,4,6-trinitrophenol
5. Benzoic acid is soluble in aqueous $NaHCO_3$ whereas phenol is not. Whereas, both dissolve in aqueous NaOH. Discuss.
6. Can you perform litmus test for water-insoluble compound in alcohol?
7. A carboxylic acid, $RC^{14}OOH$, gives off effervescence with aqueous $NaHCO_3$. Pick out the correct composition of the liberated gas:
    (a) Only CO_2
    (b) Only $C^{14}O_2$
    (c) both $C^{14}O_2$ and CO_2
8. How will you differentiate between cinnamic acid and benzoic acid by:
    (a) Physical method
    (b) Chemical method
9. How will you separate a mixture of naphthalene and benzoic acid by using:
    (a) Physical method
    (b) Chemical method
10. Benzene sulphonic acid is added to saturated solution of sodium bicarbonate. What do you think will happen?
11. How will you differentiate a monocarboxylic acid from a dicarboxylic acid in the lab.
12. Why do acids turn blue litmus red? Will phenol turn blue litmus red?
13. Why does litmus paper turn blue/red under different pH?
14. Give one chemical test to distinguish between formic acid and acetic acid.

15. What is Blanc's rule? Explain with suitable example, why glutaric acid gives flourescein test but malonic acid does not.

16. Formic acid gives test for both carbonyl group and carboxylic acids. Explain.

17. Two compounds A and B both dissolve in dilute NaOH. A is soluble in aqueous $NaHCO_3$ and gives off colourless, odourless gas. B is insoluble in aqueous $NaHCO_3$. Sodium salt of A is used as food preservative. Both A and B melt at 122°C. B forms azo dye in alkaline medium. Predict structures of A and B.

18. Which is a stronger acid and why? (A) Acetic acid and Benzoic acid (B) maleic acid/fumaric acid.

19. K_1 of oxalic acid is $>> K_2$. Explain.

20. What is the correct name for the SBT reagent and why?

21. Formation of SBT reagent is what type of reaction?

22. What is the application of SBT salt?

23. Why do you add alkali to carboxylic acid solution before adding the SBT reagent?

24. What will happen if the SBT preparation is carried out in strongly acidic/ alkaline medium?

25. What is the need to add sodium bicarbonate solution during the preparation of amide or anilide?

4.3.2.1.2 Phenols

Phenols are characterised by the presence of –OH group directly attached to the aromatic ring, as in C_6H_5OH and p-$CH_3C_6H_4OH$ and not in $C_6H_5CH_2OH$, an arylalkyl alcohol. In the industry, phenols find use as starting material in manufacture of explosives, plastics and drugs such as aspirin. Hydroquinone is a component of photographic developer, which reduces exposed silver bromide crystals to black metallic silver. Many other phenol derivatives are used in the manufacture of deeply coloured azo dyes. A mixture of cresols is used in wood preservative compositions like **creosote**.

Most phenols are solids at room temperature except m-cresol, $CH_3C_6H_4OH$, which is a liquid. All of them are colourless if pure, however they slowly turn brown due to aerial oxidation. All are characterized by an odour resembling phenol (C_6H_5OH), commonly called "**carbolic acid**". Most phenols are insoluble or sparingly soluble in water except phenol, catechol, resorcinol, and hydroquinone, which are readily soluble. All phenols are soluble in sodium hydroxide solution, in some cases (e.g., catechol, resorcinol, hydroquinone) with rapid darkening in colour.

Phenols are weaker acids than carboxylic acids and hence do not liberate carbon dioxide with aqueous sodium bicarbonate solution, even though some phenols may be soluble in it (See solubility in 5% aqueous $NaHCO_3$ *section 4.3.1.5*, page 97, solubility test). A list of commonly occurring phenols along with their pK_a values is given in Table 4.8 and Table 4.9 lists the common classification tests for phenols.

Table 4.8   Some Commonly Occurring Phenols in the Undergraduate Laboratory

S. No.	Name	pKa value**	GHS symbol*
1.	Phenol	9.99	
2.	α-Naphthol	9.34	
3.	β-Naphthol	9.51	
4.	Cresol(o-/m-/p-)	o-:10.28 m-:10.09 p-:10.26	
5.	Resorcinol	9.30 and 11.06	
6.	Hydroquinone	9.96 and 11.4	
7.	Catechol	9.45 and 12.8	

*All values are given at 25°C. Please refer to Vol. 1, Chapter 1 for details on GHS symbols

** *Source:* https://pubchem.ncbi.nlm.nih.gov

Table 4.9   Common Classification Tests for Phenols

S. No.	Classification test	Test category	Observation	Inference
1.	Solubility in 5% aq. NaOH	Preliminary test	Soluble, precipitates on acidification with 5% aqueous HCl	Acidic compound (carboxylic acid or phenol)
2.	Neutral FeCl$_3$ test	Group test	Brown, green, blue or purple colouration. (Not yellow or orange)[a]	Phenol
3.	Azo dye test	Special test	Orange or red coloured dye	Phenol
4.	Phthalein test	Special test	Red, blue or green coloration. Yellow-orange solution with green fluorescence	Phenol/cresols/ catechol Resorcinol
5.	Liebermann Nitroso test	Special test	Green or blue colouration	Phenol has Para-position free
6.	Bromine solution test	Special test	White precipitate[b]	Phenol

[a]1-/2-Naphthols give colouration only if solution is prepared in alcohol

[b]Only given by phenols with activated nucleus.

Classification tests for phenols

1. Solubility in 5% aqueous NaOH

See *section 4.3.1.5, page 95,* solubility tests, for details.

Note

- Carboxylic acids are also soluble in 5% aqueous NaOH due to formation of carboxylates.
- Perform the test for compounds insoluble in cold water only.

Safe Waste Disposal

- Carefully neutralize the test solution with 5% aqueous HCl, with stirring, until a pH ~7 is attained (use pH paper). Transfer the contents to aqueous waste container.

2. Neutral ferric chloride test: Most phenols react with neutral ferric chloride to give coloured solution (other than yellow or orange, like brown, blue, green, purple etc.) due to formation of ferric phenolate complex. The complex is formed *via* formation of coordinate covalent bond between iron(III) and phenolate ions.

$$6C_6H_5OH + FeCl_3 \longrightarrow H_3[Fe(OC_6H_5)_6] + 3\,HCl$$
Purple, blue, green or brown

Hydroquinone, however, does not form ferric phenolate. It gives green colouration due to oxidation to quinhydron. Hydroquinone is first oxidized to *p*-benzoquinone which combines with another molecule of hydroquinone to form quinhydron.

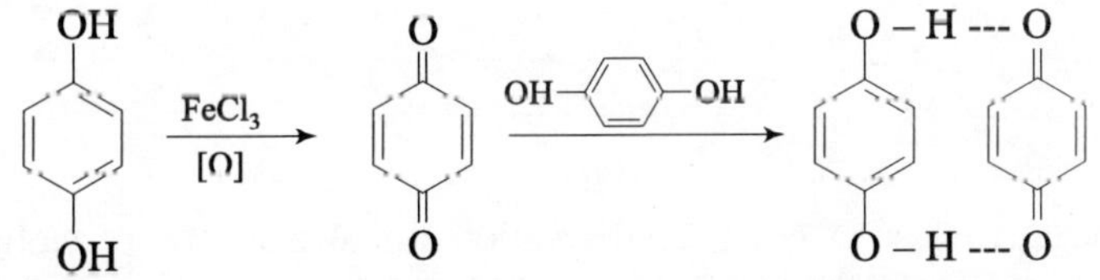

Hydroquinone   p-Benzoquinone   Quinhydron (Dark-green solid)

1-/2-Naphthols are insoluble in water, hence do not show characteristic colouration with aqueous ferric chloride. If however, alcoholic solution of 1-/2-naphthol is treated with alcoholic solution of ferric chloride, they undergo oxidative coupling to form coloured complex.

Procedure: To a very dilute solution of the unknown phenol (0.05 g in 4-5 mL water or alcohol) or to a minute crystal of the solid in a test tube, add dropwise ferric chloride solution (1% in water or alcohol depending upon the solubility of the compound). Appearance of coloured solution (green, blue, brown, violet etc.) confirms compound to be a phenol.

Note

- If compound is sparingly soluble in water, either dissolve the compound in hot water, cool, filter and use the filtrate or dissolve the compound in alcohol and add alcoholic solution of ferric chloride.
- Some amines, enolisable aldehyde, ketones and compounds like ethyl acetoacetate also give this characteristic test.
- Acidic phenols like like *p*-hydroxybenzoic acid and nitrophenols do not give characteristic colouration. However, salicylic acid and salicylaldehyde gives violet colour due to chelate formation.
- A negative test is not a confirmation for absence of phenolic group. Please correlate the solubility test, Baeyer's reagent test and bromine solution test before arriving at a conclusion.
- Neutral salts of carboxylic acids give buff coloured precipitate with ferric chloride. The precipitate however dissolves in dilute hydrochloric acid.
- Aqueous solution of ferric chloride kept in the lab is acidic due to hydrolysis. Hence, the solution should be neutralized (see Appendix II) before use or a freshly prepared solution of ferric chloride should be used.

Safe Waste Disposal

- Transfer all the aqueous solutions to hazardous waste containers.

3. Azo dye test: This test is based on the coupling reaction of phenols with diazonium salts in alkaline pH (9-10). See the discussion on amines, *section 4.3.2.5*, page 209, for details.

Procedure: Take aniline (2-3 drops) in a clean test tube, add concentrated hydrochloric acid (1.0 mL) and then add water (3.0 mL), dropwise, to dissolve any precipitated aniline hydrochloride. Cool the solution in ice. Add cold, aqueous sodium nitrite solution (2-3 drops, 20%). Allow the solution to stand in ice for 4-5 minutes to facilitate complete diazotization. Add this cold diazonium salt solution to a pre-cooled solution of phenol in excess of aqueous NaOH solution (5.0 mL, 10%). Coloured precipitate of azo dyes ranging from orange to dark red is obtained, depending upon the nature of phenol used.

Note

- Ensure that the temperature of the reaction medium is maintained between 0-5°C.
- Catechol decomposes under the reaction conditions.

Safe Waste Disposal

- Separate precipitate by filtration and transfer organic wastes to non-hazardous solid waste container. Place the remaining liquid in aqueous waste container after neutralizing with 5% aqueous hydrochloric acid to pH ~7.

4. Phthalein test: Certain phenols combine with phthalic acid or phthalic anhydride in the presence of concentrated sulphuric acid to form phthalein/xanthine dyes, which give characteristic colouration in the alkaline medium.

Note

- Resorcinol under similar conditions forms fluorescein, a xanthine dye. (see *section 4.3.2.1.1,* page 127, carboxylic acids for details).
- The pink colour of phenolphthalein disappears on adding excess of alkali.

The characteristic colouration with different phenols has been shown in Table 4.10.

Table 4.10   Characteristic Colouration of Common Phenols in Phthalein Test

S. No.	Phenol	Colouration
1.	o-Cresol or Phenol	Red
2.	m-Cresol	Bluish purple
3.	p-Cresol	None
4.	Catechol	Blue (Alizarin)
5.	Resorcinol	Yellow-Orange solution with green fluorescence
6.	Hydroquinone	Purple
7.	1-Naphthol	Green
8.	2-Naphthol	Faint green with some fluorescence

Procedure: See fluorescein test, *section 4.3.2.1.1* page 127, and follow the same procedure. Only replace the carboxylic acid by phthalic anhydride and resorcinol by the organic compound.

Safe Waste Disposal

Place all test solutions in aqueous waste container after neutralizing to pH ~7 with 5% aqueous hydrochloric acid and run down the sink.

5. Liebermann Nitroso Test: Phenols (except *p*-substituted phenols, nitrophenols and catechol) combine with nitrous acid to form *p*-nitrosophenol which further reacts with the second molecule of phenol to form indophenol (an acid-base indicator). Indophenol gives green colouration in acidic medium and blue coloration in alkaline medium.

Procedure: To one minute crystal of sodium nitrite taken in a clean dry test tube, add unknown phenol (0.5 g) and heat very gently on a Bunsen burner flame for about 20 seconds to allow the contents to change to molten state. Allow the contents to cool and add dropwise concentrated sulphuric acid (2-3 drops). Rotate the tube gently to mix the contents. A deep green or deep blue coloration develops slowly. Dilute cautiously with water; observe that the solution turns red. Now add an excess of sodium hydroxide solution (10%), the solution turns blue or green again. A positive test confirms that phenol has para-position free.

$$2NaNO_2 + H_2SO_4 \longrightarrow 2HNO_2 + Na_2SO_4$$

Phenol $-OH \xrightarrow{HO-N=O} HO-\langle\rangle-N=O \underset{\text{Tautomerism}}{\rightleftharpoons} O=\langle\rangle=N-OH$ (Yellow)

(1) $HO-\langle\rangle$
(2) H_2SO_4

$$O=\langle\rangle=N-\langle\rangle-OH \xleftarrow{H_2O\,(excess)} \left[\overset{\oplus}{HO}=\langle\rangle=N-\langle\rangle-OH\right] HSO_4^{\ominus}$$

Indophenol (red) — (Blue/Green)

NaOH

$$O=\langle\rangle=N-\langle\rangle-O^{\ominus}Na^{\oplus}$$

Sodium salt of indophenol (deep blue/green)

Note

- Precise colour varies with the purity of the sample, amount of reagent used, temperature and duration of heating.
- This test is unsatisfactory with many phenols like nitrophenols, catechols and hydroquinone.

Safe Waste Disposal

- Neutralize the aqueous solution with 5% aqueous hydrochloric acid until pH ~7 and then transfer into aqueous solution container.

6. Bromine solution test: See the discussion on unsaturation test, *section 4.3.1.8,* page 117, for details on theory of this test. Many phenols, except those with strong reducing groups, undergo ring substitution with bromine in water/acetic acid/CCl_4 to form a white precipitate of 2,4,6-tribromophenol. The reaction leads to decolourisation of bromine water with evolution of HBr gas, if the solvent is not water. Rate of reaction is faster in polar solvents like water than in CCl_4.

$$\text{Phenol} \quad + 3Br_2 \xrightarrow{H_2O} \text{2,4,6-Tribromophenol} \quad + 3HBr$$

(White ppt.)

Procedure: To a concentrated aqueous solution of the given phenol [(0.25 g) in water or dilute HCl (10.0 mL)] add dropwise saturated solution of bromine in water. Initially, the bromine solution is decolorised and then on adding an excess of reagent, a white or yellowish-white precipitate of a tribromo derivative is produced.

Note

- Aromatic amines and unsaturated hydrocarbons also decolorize bromine solution. Correlate your data carefully.
- All phenols except catechol, hydroquinone, 1- and 2-naphthol give this test.
- Catechol gives a deep red colouration immediately on addition of one drop of bromine water.
- Hydroquinone produces a deep red coloured solution on gradually adding bromine water. Gradually, deep green crystals precipitate, which dissolve in solution to produce yellow colouration.
- 1- and 2-Naphthol decolourize bromine solution, but usually no precipitate of the bromo compound can be obtained.
- The test is not suitable for water insoluble phenols, due to the problem in distinguishing the unknown phenol from the bromo derivative. Aqueous alcohol may be used as solvent in such cases.
- No evolution of HBr is seen if water is used as solvent because HBr is soluble in water.

Safe Waste Disposal

- Place the test solution in the container for halogenated waste.

7. Spot test for phenols

(a) Neutral ferric chloride test: Prepare concentrated solution of the compound and take freshly prepared concentrated solution of ferric chloride. Spot the solutions one over the other on a Whatman paper strip as given below:

- First spot the concentrated solution of the compound.
- Then introduce the freshly prepared concentrated solution of ferric chloride over it.

A dark coloured spot (brown, blue, green, violet etc.) confirms a phenol.

(b) Liebermann Nitroso test: Prepare concentrated solution of your compound, sodium nitrite and a dilute solution of sodium hydroxide. Spot the solutions one over the other on a TLC plate or Merck TLC strip as given below:

- First spot concentrated solution of the compound.
- Then introduce concentrated aqueous solution of sodium nitrite.

- Spot concentrated sulphuric acid (two times)
- Heat the plate in an oven maintained at 120°C for one minute. Ensure that the spot has dried completely.
- Take out the strip and introduce phenol once more.
- Heat again in the oven for one minute.
- Scratch the spotted silica gel and transfer to a test tube containing dilute sodium hydroxide solution (2-3 mL).

Development of blue colour confirms that the phenol has para position free.

(c) Phthalein Test: Prepare concentrated solutions of your compound, sodium hydroxide, phthalic acid/phthalic anhydride, and spot them one over the other two times each on a TLC/Merck plate, in the sequence mentioned below:

- First spot the concentrated aqueous solution of phthalic acid/phthalic anhydride.
- Next, spot the concentrated solution of the given phenol.
- Next introduce 2 drops of concentrated sulphuric acid.
- Heat the plate in oven (maintained at ~120°C) for 40-50 seconds, till the spots dry up completely. Scratch the spotted portion of the gel and introduce it into a test tube containing very dilute aqueous sodium hydroxide solution (5-7 drops)

Red, blue, green or purple coloured spot confirms phenol

8. Spectroscopic identification of phenols

Table 4.11 Spectral Data of Phenols

S. No.	Type of spectrum	Peak value	Peak attribution	Remarks
1.	IR	3650-3600 cm^{-1}	O-H str. (Free)	Strong, sharp
		3300-3200 cm^{-1}	O-H str. (H-bonded)	Medium, broad
		1260-1000 cm^{-1}	C-O str.	
		>3000 cm^{-1}	Aromatic C-H str.	
2.	NMR	4-12 δ*	Ar-OH	Replaceable with D_2O
		7 δ	Ar-H	Position and split pattern depend upon the position & number of substituents in the ring.

*δ value depends upon the concentration of the compound and temperature.

Derivatives of phenols

1. Benzoate: Phenols can be converted into water-insoluble benzoates *via* **Schotten-Baumann reaction,** which involves replacement of active hydrogen present in phenols with benzoyl chloride under the influence of aqueous sodium hydroxide. For details, see benzoylation, *section 6.8*, volume 1.

OH   O=C–Cl

+   →   Aqueous NaOH   →   COOC$_6$H$_5$ +   NaCl + H$_2$O

Phenol   **Benzoyl chloride**   **Phenyl benzoate**

Procedure: Place phenol (1.0 g) and 10% aqueous NaOH (15.0 mL) in a conical flask fitted with a stopper. Add one lot of benzoyl chloride (1.5 mL, in two lots of 0.75 mL each) and stopper the flask immediately. Shake the reaction mixture vigorously for 10-15 minutes or till contents come back to room temperature, then add the second lot of benzoyl chloride and repeat as stated before. Allow the contents to come back to room temperature. Filter the separated solid on suction pump, wash well with water. Dry the solid and recrystallize a fraction of the crude sample using ethanol. Filter the crystallized solid and dry on a porous clay plate and record its melting point.

Note

- If lump formation occurs, add a few pellets of sodium hydroxide to the conical flask, shake and allow to stand for 15-20 minutes. This will hydrolyze unreacted benzoyl chloride.
- Carefully adjust the reagent amount for polyhydroxy compounds.
- NaOH is used in the reaction to: (a) remove HCl and push equilibrium towards right, (b) hydrolyze unreacted benzoyl chloride, (c) accelerate reaction by converting phenol to phenoxide ion (See Volume 1, *section 6.8*, benzoylation, for details).
- Before working up ensure that the medium is alkaline and does not smell of benzoyl chloride.

Safe Waste Disposal

- To the filtrate, add 5% aqueous HCl, dropwise, until pH ~ 7. Filter any precipitate and discard it in organic waste container. Run the aqueous layer down the drain.

2. 3,5-Dinitrobenzoate: Phenols are known to form crystalline 3,5-dinitrobenzoates on reaction with 3,5-dinitrobenzoyl chloride in the alkaline conditions.

Phenol   3,5-Dinitrobenzoyl chloride   **Phenyl-3,5-dinitrobenzoate**

Procedure: Dissolve phenol (1.0 g) in 10% sodium hydroxide (20.0 mL) in a dry mortar. Then add 3,5-dinitrobenzoyl chloride (0.5 g, **see Appendix II**). Grind the contents for 1-2 minutes. Transfer to a beaker and warm the contents on steam bath, in a fume hood. When the reaction progresses, the contents will liquify. Continue heating on steam water bath for another 10-15 minutes. Transfer the contents with stirring to a beaker containing ice-water (50 mL). Filter the precipitated solid, wash with saturated solution of sodium bicarbonate and then with water. Recrystallize a part of the crude sample using ethanol. Filter the crystallized solid and dry on a porous clay plate and record its melting point.

Note

- 3,5-Dinitrobenzoyl chloride is a strong irritant and moisture sensitive. Always wear gloves while handling.
- Ensure completion of reaction before work-up.
- Washing with sodium bicarbonate removes any unreacted 3,5-dinitrobenzoic acid.

Safe Waste Disposal

- Neutralize the alkaline solution with 5% aqueous HCl. Filter and transfer the organic solid to organic waste container. Run the neutral solution down the drain.

3. Acetate: Phenols like 2-naphthol, hydroquinone, salicylic acid, vanillin, etc., are known to undergo acetylation with acetic anhydride, in alkaline conditions, due to presence of active hydrogen in them. For example, hydroquinone forms hydroquinone diacetate.

Hydroquinone   Acetic anhydride   **Hydroquinone diacetate/**
1,4-Diacetoxy benzene

Vanillin   **Acetic anhydride**   **4-Acetoxy-3-methoxy benzaldehyde**

Procedure: In a clean conical flask (100 mL) fitted with a stopper, dissolve the given phenol (1.0 g) in 10% aqueous NaOH solution (15.0 mL). Add crushed ice, followed by addition of acetic anhydride (1.5 mL), dropwise with stirring. Loosely stopper the flask and shake the contents for 15-20 minutes. Filter the separated solid on suction pump, wash well with water. Dry the solid and recrystallize a fraction of the crude sample using hot water or aqueous ethanol. Filter the crystallized solid, dry on a porous clay plate and record its melting point.

Note

- If no solid separates on shaking (15-20 minutes), acidify with 10% sulphuric acid.
- See Volume 1, *section 6.7*, acetylation for detailed theory.
- Acetates of monohydric phenols are usually liquids but those of di- or trihydric phenols and many other substituted phenols are crystalline solid.

Safe Waste Disposal

- Neutralize aqueous solution with 5% aqueous HCl until pH ~7. Transfer to aqueous waste container.

4. Bromo derivative: Phenol with free *o/p*-position reacts rapidly with bromine water to give bromo derivative with bromine substitution at *o/p*-position. The reaction proceeds *via* electrophilic substitution on the ring.

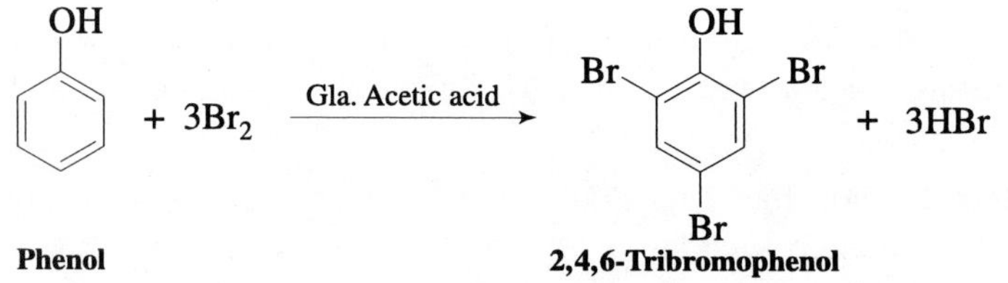

OH + 3Br$_2$ $\xrightarrow{\text{Gla. Acetic acid}}$ 2,4,6-Tribromophenol (with Br at 2,4,6 positions) + 3HBr

Phenol → 2,4,6-Tribromophenol

Procedure: To a clean conical flask (250 mL) containing solution of the given phenol (1.0 mL/1.0 g) in glacial acetic acid (4.0 mL), add a solution of bromine in acetic acid (7% v/v, 25.0 mL), dropwise with constant shaking. Cover the conical flask with filter paper and allow the reaction mixture to stand for 20 minutes. The reaction mixture acquires a faint orange-yellow colour due to the presence of excess of bromine. After 20 minutes, transfer the contents to a beaker containing ice-cold water. Filter the separated solid on suction pump, wash well with water. Dry the solid and recrystallize a part of crude sample using dilute ethanol. Filter the crystallized solid, dry on a porous clay plate and record its melting point.

Note

- Avoid bromination, if possible, as bromine is corrosive.
- Any unreacted bromine may be removed by adding sodium bisulphite in mild acidic medium. (See Volume I, *section 6.3.1*, bromination)
- Bromination of easily oxidizable phenols like hydroquinone is difficult due to partial oxidation by bromine.

Safe Waste Disposal

- Transfer all the contents to halogenated waste container.

Post Lab Questions

1. Salicylic acid gives $FeCl_3$ test whereas *p*-hydroxy benzoic acid does not. Explain.
2. Quinol does not give Liebermann's test. Explain.
3. *p*-Nitrophenol is yellow in water but deep yellow in alkali. Explain.
4. Picric acid does not give $FeCl_3$ test. Explain.
5. Bromination of phenol is faster in water than CCl_4. Explain.
6. Why do you need to use neutral $FeCl_3$ solution for testing phenols?
7. How will you neutralise $FeCl_3$ solution?
8. Does 2-naphthol react with aqueous $FeCl_3$. How can you modify this test to get colouration with 2-naphthol?
9. Bromination of phenol in bromine water is different from bromination of styrene. Explain.
10. Give a chemical test to distinguish between the following pairs of phenols:
    (a) p-Cresol and o-Cresol
    (b) Phenol and Resorcinol
    (c) Phenol and *p*-nitrophenol
11. Phenols give azo dye test in the alkaline medium. Explain.
12. Pink colour in phthalein test for phenol disappears if excess of alkali is added. Explain.

4.3.2.2 *Compounds Containing Carbon, Hydrogen and Oxygen (Neutral Compounds)*

Many organic compounds containing oxygen and/or nitrogen are neutral. In this section we are first taking neutral compounds containing carbon, hydrogen and oxygen. These contain the following four functional groups: (a) aldehydes and ketones, (b) esters, (c) alcohols, and (d) carbohydrates. This entire group of compounds with the exception of some lower members are insoluble in water and dissolve in concentrated sulphuric acid due to formation of oxonium salts (Note: Simple sugars are water soluble). They should be tested for the functional group in the sequence described else the reactions may lead to erroneous conclusions.

Note

- Alkenes and some aromatic hydrocarbons are also soluble in concentrated sulphuric acid. Please refer *section 4.3.1.5*, page 98, for solubility in concentrated sulphuric acid.

4.3.2.2.1 *Aldehydes and Ketones*

These are compounds characterised by the presence of carbonyl group ($>C=O$) and are therefore called **carbonyl compounds**. Other compounds like esters, carboxylic acids, anhydrides and amides, which also contain carbonyl group, are not called carbonyl compounds as properties of these compounds are quite different from those of aldehydes and ketones.

Most of the aldehydes and ketones are colourless liquids (except formaldehyde, which is a gas) with pungent odour. For example, benzaldehyde has a bitter almond smell, cinnamaldehyde has a cinnamon smell and salicylaldehyde smells like phenol. These compounds are mostly insoluble in water (except the lower members containing up to four carbon) however, they dissolve readily in alcohol, ether and concentrated sulphuric acid. Various aldehydes and ketones find commercial applications. For example, formalin is used to store biological samples because it is resisitant to bacterial growth, acetone is a key ingredient in nail polish removers and is a solvent in paints. Benzaldehyde is found in almonds and cinnamaldehyde in cinnamon etc.

Aldehydes and ketones are characterised in the lab through their behaviour towards oxidising agents and through their reaction with ammonia derivatives like: 2,4-dinitrophenylhydrazine, phenylhydrazine hydrochloride, hydroxylamine hydrochloride, semicarbazide hydrochloride, etc. The reactivity of individual compound towards these reagents varies with the nature of substituent attached to the carbonyl group. However, as a general rule, aldehydes are more reactive than ketones towards these reagents. The reaction with ammonia derivatives involves nucleophilic addition-elimination reaction. Table 4.12 lists the classification tests for aldehydes and ketones.

Table 4.12   Classification Tests for Aldehydes and Ketones

S. No.	Classification test	Test category	Observation	Inference
1.	Solubility in concentrated sulphuric acid	Preliminary test	Soluble due to formation of oxonium salts[a].	Oxygen containing compound or unsaturated hydrocarbon or aromatic compound like anthracene
2.	2,4-DNP test	Group test	Yellow-orange precipitate insoluble in water[b]	Carbonyl compound

3.	Tollens' test	Special test	Silver mirror on the walls of test tube[c]	Aldehyde, if –ve then ketone
4.	Fehling solution test[d]	Special test	Brick red precipitate	Aliphatic aldehyde
5.	Schiff's test[d]	Special test	Purple colouration	Aldehyde
6.	Iodofom test[e]	Special test	Yellow precipitate	Methyl ketone
7.	Nitroprusside test[e]	Special test	Red colouration	Methyl ketone

a: unsaturated hydrocarbons and some arenes are also soluble; **b:** Amines form hydrochlorides which are water soluble; **c:** Some aliphatic methyl ketones may give black precipitate; **d:** to be done only if Tollen's test is +ve; **e:** to be done if DNP test is +ve and Tollens' test is either –ve or a black residue is formed.

Classification tests for Aldehydes and Ketones

1. Solubility in concentrated sulphuric acid: See *section 4.3.1.5,* page 98, solubility in concentrated sulphuric acid.

2. 2,4-Dinitrophenylhydrazine (2,4-DNP) test: The reaction of aldehydes and ketones with 2,4-dinitrophenylhydrazine to form partially soluble 2,4-dinitrophenylhydrazone is probably the most studied and most successful of all qualitative tests and derivatizing procedures for aldehydes and ketones. The reaction involves nucleophilic addition-elimination reaction to form the coloured 2,4-dinitrophenylhydrazone. Nucleophilic 2,4-DNP attacks the electrophilic carbon of the carbonyl group, in the presence of mineral acid as catalyst, to from tetrahedral addition product which readily undergoes dehydration in the reaction conditions to form the desired imine as product. The colour of the 2,4-dinitrophenylhydrazone depends on the nature of carbon skeleton attached to the carbonyl group. The aliphatic compounds give a yellow coloured product however, conjugation to C=C or aromatic ring shifts the colour to orange-red, due to extended conjugation.

2,4-Dinitrophenylhydrazine   **Carbonyl Compound** (Aldehyde/ketone)   **2,4-Dinitrophenylhydrazone of carbonyl compound**

Note: For detailed theory, see Volume I, *section 6.12.1.4, Addition-Elimination Reactions of Carbonyl compounds.*

Procedure: *For water-soluble compounds:* Dissolve the unknown compound (0.1 g or 0.2 mL, if liquid) in water (2-3 mL). Add 2,4-DNP in water (5-7 mL, see Appendix II). Allow the contents to stand for some time. If precipitation does not start, scratch the side walls with glass rod. If no change is observed, heat on a boiling water bath for 5-10 minutes and then allow the reaction mixture to stand at room temperature for 10-15 minutes. Cool in ice, if required. A yellow precipitate of 2,4 dinitrophenylhydrazone indicates the presence of carbonyl group.

For water-insoluble compounds: Dissolve the given organic compound (0.2 mL) in methanol/ethanol (2-3 mL) and then add 2,4-DNP in alcohol (2.0 mL, see Appendix II). If no precipitate is formed in cold, warm the contents on boiling water bath for 5-10 minutes and cool. Orange-red coloured precipitate confirms the presence of a carbonyl group.

Note

- Brady's reagent (aqueous-alcoholic solution of 2,4-DNP) can be used to test for both water soluble and water insoluble compounds (see Appendix II for details about preparation).

- The alcohol used for dissolving the compound should be free from aldehyde/ketone as impurity. Otherwise some turbidity may be obtained on adding 2,4-DNP, even in the absence of carbonyl compound. It is therefore advisable to carry out a blank test whenever there is some doubt.

- For some aliphatic ketones (like acetone), dropwise addition of the solution of compound to the reagent works better.

- Mineral acids, like HCl or H_2SO_4 present in the reagent is sufficient to catalyse the addition-elimination reaction, which occurs in a mildly acidic pH range (~4-5).

- In some cases, 2,4-dinitrophenylhydrazine itself may get precipitated as an orange colour solid. However, it dissolves in dilute sulphuric acid whereas 2,4-dinitrophenylhydrazone does not.

- 2,4-DNP is a mild irritant and stains the skin. Use gloves while performing this test.

- Aromatic amines, like aniline, precipitate as water soluble salt with 2,4-DNP reagent due to presence of mineral acid.

Safe Waste Disposal

- Place the test solution in the hazardous organic waste container.

Distinguishing tests for Aldehydes and Ketones

Aldehydes being more reactive than ketones can be easily oxidised by mild oxidising agents like Tollens' reagent, Fehling solution, Schiff reagent and Benedict solution. Ketones are oxidised under drastic conditions. Methyl ketones can however be oxidised by sodium hypobromide in the alkaline medium (Iodoform test).

Note

- These tests should be performed only if 2,4-DNP test is +ve.

3. Tollens' test: Tollens' reagent is ammonical solution of silver nitrate. It easily oxidises aldehydes to corresponding acid salts and is itself reduced to metallic silver, which can be seen as a black coloured precipitate or a silver mirror on the inner walls of the test tube.

$$AgNO_3 + NaOH\ (aq.) \longrightarrow Ag_2O\downarrow \xrightarrow{NH_4OH} [AgNH_3)_2]^+\ OH^-$$

$$\underset{\textbf{Aldehdye}}{RCHO} + \underset{\textbf{Tollens reagent}}{2[Ag(NH_3)_2]OH} \longrightarrow RCOONH_4 + 3NH_3 + H_2O + 2Ag\downarrow$$

Procedure: Take Tollens' reagent (2.0 mL) in a clean test tube (cleaned with dilute nitric acid) and add the unknown compound (2-3 drops), if no silver mirror is obtained in cold, put the test tube in a beaker containing water and heat on a wire gauze using Bunsen burner. Appearance of silver mirror confirms the presence of an aldehyde.

Note

- Wash the residue immediately with dilute nitric acid to dissolve metallic silver, deposited on the inner walls of the test tube.

- Use freshly prepared Tollens' reagent only (See **Appendix II** for preparing the reagent).

- Reducing sugars like glucose, fructose, some phenols, quinones, α-hydroxyketones some formates, etc., also reduce Tollens' reagent.

- Water insoluble aldehydes may react slowly. In such cases, prepare a solution of the given compound (2-3 drops) in aldehyde-free alcohol (2-3 mL) and use.

- Sometimes metallic silver is precipitated as a grey or black precipitate, especially when the glass test tube is not thoroughly cleaned.

- This reaction is autocatalysed by silver metal and hence requires induction period of a few minutes. The test solution decomposes on standing and deposits a highly explosive precipitate of silver fulminate, AgCNO (please see, disposal).

- The reaction mixture should be warmed and not boiled after adding Tollens' reagent, as boiling may lead to false positive due to decomposition of the reagent.

Safe Waste Disposal

- Pour the test solution into a beaker. Add 5% nitric acid dropwise until the metallic silver dissolves completely. Neutralize the solution by adding solid sodium carbonate. Next add saturated sodium chloride solution to it until all the silver in the solution is precipitated as silver chloride. Filter the solution and place the residue of silver chloride in the non-hazardous solid waste container. Transfer the filtrate to the aqueous waste solution container.

4. Schiff's test: This test is based on the fact that aliphatic aldehydes immediately convert the colourless Schiff's reagent to magenta pink whereas ketones do not react. (**Note:** Benzaldehyde restores the pink colour on standing for some time.)

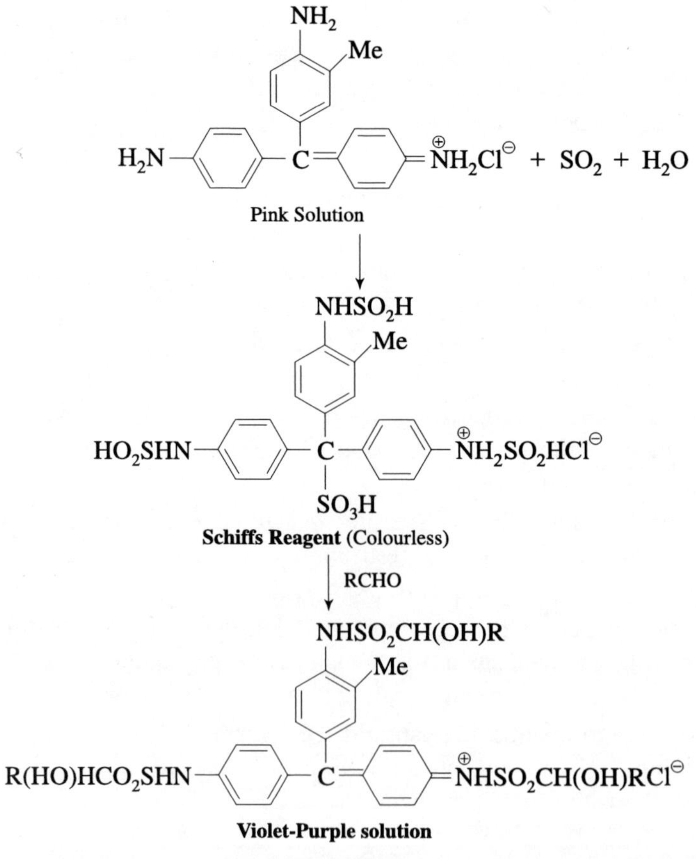

Fuchsin dye or Pararosaniline hydrochloride is a triarylmethane dye. Schiff's reagent is prepared by passing sulphur dioxide gas through pink coloured aqueous acidic solution of pararosaniline hydrochloride, when a virtually colourless Schiff's reagent is formed. The reaction with sulphur dioxide involves conversion of Fuchsin dye to the colourless leucosulfonic acid by sulphurous acid. The reaction apparently involves addition of sulfurous acid to the quinoid nucleus of the dye thereby converting it into colourless benzenoid form. The leucosulphonic acid is unstable and loses sulphurous acid on reaction with aldehydes to generate magenta (not pink!) coloured solution.

Procedure: Take the given organic compound (2-3 drops) in a test tube and add Schiff's reagent (1-2 mL). An instant magenta colour develops on shaking in the cold.

Note

- Benzaldehyde restores the colour very slowly.
- Do not heat the contents while performing the reaction because pink colour will develop even in absence of aldehyde.
- Exposure to air, alkali and alkali metal salts of weak acids restore the pink colour of Fuschin dye.
- Some ketones and unsaturated compounds react with sulphurous acid to regenerate pink colour of Fuchsin (different from magenta colour produced by aldehydes). So, conclusion should be drawn with caution.

Safe Waste Disposal

- Neutralize the test solution to pH ~7 with sodium carbonate and place in the aqueous waste solution container.

5. Fehling's solution test: This test is used for the distinction between aliphatic and aromatic aldehydes. Aliphatic aldehydes being more reactive, reduce the cupric ions in the Fehling's solution to produce red coloured cuprous oxide.

Fehling's solution contains intense blue coloured soluble complex of copper tartarate in the strongly alkaline medium, formed by mixing equal volumes of aqueous copper sulphate (Fehling A) and alkaline solution of Rochelle's salt (Fehling B). It is not stable, as copper hydroxide gets precipitated on standing. Copper hydroxide is a poor oxidising agent, primarily because of its low solubility in aqueous alkaline medium. Formation of soluble complex of copper tatarte (Fehling's solution) or copper citrate (Benedict's reagent) increases its solubility and thus the oxidation rate. When heated with aliphatic aldehyde, Fehling's solution oxidises aldehyde to corresponding acid salt and is itself reduced to red coloured cuprous oxide.

$$CuSO_4 \; + \; 2NaOH \longrightarrow Cu(OH)_2 \; + \; Na_2SO_4$$

Copper complex with sodium potassium tartrate (soluble)

$$RCHO \; + \; 2Cu^{2\oplus} + \; 5OH^{\ominus} \longrightarrow Cu_2O\downarrow + RCOO^{\ominus} + \; 3H_2O$$

Brick red ppt.

This reaction is common for Fehling's solution and Benedict's reagent (Also see page 155)

Procedure: To aldehyde (2-3 drops) taken in a clean test tube, add 10% aqueous Na_2CO_3 solution (1.0 mL, *See Note below*) and then add Fehling's solution (2-3 mL, prepared by mixing equal volumes of Fehling A and B). Heat the mixture on boiling water bath for 3-4 minutes. The solution usually turns green and on standing brick red precipitate of cuprous oxide slowly separates.

Note

- Aliphatic aldehydes usually fail to give Fehling's solution under normal conditions, as aerial oxidation of aliphatic aldehydes makes their aqueous solutions acidic. The base in Fehling's solution is neutralised by the aliphatic acid formed. Use of an excess of Fehling's solution is not suggested as the excessive blue colour may mask the subsequent reduction. Therefore, it is advisable to neutralise the aqueous solution of aliphatic aldehydes with sodium carbonate before adding Fehling's solution.

- Reducing sugars and α-hydroxyketones also give positive Fehling's test.

- Fehling's solution is not stable and decomposes on standing due to precipitation of copper hydroxide. Therefore, it should be prepared fresh by mixing equal volumes of Fehling A and B. Both of which are highly stable at room temperature.

- A blank experiment should be performed to ensure that no reduction takes place on boiling.

- Color of cuprous oxide(red or yellow) is dependant on the size of the particles, rate of reduction and concentration of solution.

Safe Waste Disposal

- Isolate the cuprous oxide precipitate and transfer it to non-hazardous solid waste container. Place the filtrate in aqueous solution container.

6. Benedict's reagent test: Benedict's reagent is a modification of Fehling's solution. It has the following advantages over Fehling's solution:

(a) More stable

(b) Involves use of single solution. It is an alkaline solution containing copper sulphate and sodium citrate. Formation of copper citrate complex decreases concentration of Cu^{2+} ions in the solution, thus preventing precipitation of cupric hydroxide.

Procedure: Take the given organic compound (2-3 drops) in a test tube and add Benedict's reagent (2.0 mL). Then heat on a boiling water bath for 5 minutes. Brick red precipitate of cuprous oxide slowly separates out in case of aliphatic aldehydes.

Note

- α-Hydroxy ketones, α-keto aldehydes and reducing sugars also reduce Benedict's reagent.

Safe Waste Disposal

- Isolate the cuprous oxide precipitate and transfer it to non-hazardous solid waste container. Place the filtrate in aqueous solution container.

7. Iodoform test: The iodoform test is a characteristic test for methyl ketones and secondary alcohols having CH_3CHOH–group. Both these types of compounds are oxidised with an alkaline solution of iodine to produce a canary yellow precipitate of iodoform.

$$R-\underset{\underset{H}{|}}{\overset{\overset{OH}{|}}{C}}-CH_3 + I_2 + 2OH^{\ominus} \longrightarrow R-\overset{\overset{O}{\|}}{C}-CH_3 + 2I^{\ominus} + 2H_2O$$

$$\underset{\substack{\text{\textbf{α-Methyl Ketone/}}\\\text{\textbf{Acetaldehyde}}}}{R-\overset{\overset{O}{\|}}{C}-CH_3} + 3I_2 + 4OH^{\ominus} \longrightarrow R-\overset{\overset{O}{\|}}{C}-O^{\ominus} + 3I^{\ominus} + 3H_2O + \underset{\substack{\textbf{Iodoform}\\\text{(Yellow ppt.)}}}{CHI_3\downarrow}$$

The mechanism of the reaction involves formation of a succession of enolate ions, which are finally iodinated. In the final steps, hydroxide displaces the triiodomethyl anion through an addition-elimination pathway. For details, see Volume 1, *section 6.11.4.*

Procedure: Dissolve the given organic compound (2-3 drops or 0.1g) in water (2.0 mL) taken in a clean test tube. Add sodium hydroxide solution (1.0 mL, 10%), and then dropwise, with shaking, add iodine-potassium iodide solution until a definite dark brown colour of iodine persists. Put a glass rod and warm the test tube in a water bath maintained at 60°C for 5 minutes. Remove the test tube from water bath and add dropwise, with shaking, aqueous sodium hydroxide solution (10%) to discharge any residual iodine (brown colour) in the solution. A canary yellow precipitate of iodoform confirms the presence of methyl ketone or secondary alcohol of the type CH_3CHOH-.

Note

- Add more iodine–potassium iodide reagent, if needed, to maintain the brown color of the solution, while 5 minutes of heating.
- Acetaldehyde is the only aldehyde and ethanol is the only primary alcohol to give an iodoform test.
- Use dioxane or methanol to dissolve the compound, if it is water insoluble.
- In case iodoform is reddish brown in colour, add more of 10% aqueous sodium hydroxide solution until brown colour of the iodine is discharged.

Safe Waste Disposal

- Place all test solutions in a beaker. Any residual iodine is removed by adding a few drops of acetone. (Ensure that the solution becomes colorless.) Filter and transfer the residual iodoform to halogenated-waste container. Add 5% aqueous HCl dropwise to the filtrate until neutral to pH paper. Run the filtrate down the drain or pour it into the aqueous inorganic waste container.

8. Sodium Nitroprusside test: Methyl ketones react with sodium nitroprusside, in the presence of excess of alkali to form reddish coloured complex. The colouration is caused by the conversion of acetone to the $CH_3COCH_2^-$ ion, which then reacts with the nitroprusside ion $[Fe(CN)_6NO]^{2-}$ giving the highly coloured ion, $[Fe(CN)_5NO(CH_2COCH_3)]^{2-}$.

Procedure: Add freshly prepared solution of sodium nitroprusside (1.0 mL) to a test tube containing unknown compound (0.5 mL). Add excess of dilute sodium hydroxide solution dropwise, with shaking. A red coloration confirms methyl ketone.

Safe Waste Disposal

- After filtration, neutralize the aqueous test solution with 5% aqueous HCl and transfer to aqueous waste solution container.

9. Spot tests for aldehydes and ketones

(a) *2,4-DNP test:* Prepare concentrated aqueous solution of your compound and 2,4-DNP reagent and spot them on TLC plate or Merck strip, as detailed below:

- Spot the liquid compound or its concentrated solution.
- Then spot alcoholic solution of 2,4-DNP

Appearance of orange/red spot confirms presence of carbonyl group.

Note

- This test works well for aromatic aldehydes and ketones only.
- Blank test should be performed simultaneously by spotting 2,4-DNP reagent on TLC plate for comparison of the spot colour.

(b) *Fehling's solution test:* Spot the following on TLC plate or Merck strip one over the other, as detailed below:

- First spot the liquid compound or its concentrated alcoholic solution.
- Concentrated Fehling A (1-2 drops)
- Concentrated Fehling B (1-2 drops)
- Heat in oven maintained at ~120°C for 1-2 minutes.

Development of red spot confirms the presence of aliphatic aldehyde.

(c) *Schiff's test:* Take a thin strip of Whatman filter paper and spot the following solutions one over the other, in the order given below:

- Concentrated solution of the compound or the liquid compound
- Schiff's reagent (from the kit)

Instant development of a magenta coloured spot indicates the presence of an aliphatic aldehyde.

Note

- Only instant pink colour is taken as positive

10. Spectroscopic identification of aldehydes and ketones

Table 4.13   Spectral Data of Aldehydes and Ketones

S. No.	Type of spectrum	Peak value	Peak attribution	Remarks
1.	IR	1715-1700 cm^{-1}	>C=O str.(Ketones)	Strong, narrow
		1740-1725 cm^{-1}	>C=O str*(Aldehydes)	Strong, narrow
		2830 & 2720 cm^{-1}	-C$\underline{H}$O	Doublet,
		1400 cm^{-1}	C$\underline{H_3}$CO-.	C-H bending
2.	NMR	9-10 δ**	—C$\underline{H}$O	Coupling of C$\underline{H}$O with α-hydrogen is observed, if present.
		2-2.5 δ	—C$\underline{H}$CO—	

* Aldehydes absorb at slightly higher frequency than ketones.

** Only formyl and aldehydic protons appear in this region.

Derivatives of aldehydes and ketones: 2,4-Dinitrophenylhydrazones, semicarbazones, oximes, phenylhydrazones, etc., are the most commonly prepared derivatives of carbonyl compounds. Iodoform derivative can also be prepared for aldehydes/ketones which contain CH_3-CO group. Oxidation of the carbonyl compounds with alkaline potassium permanganate leading to the formation of carboxylic acids can also be carried out. This reaction is especially useful for aromatic compounds.

1. 2,4-Dinitrophenylhydrazone: 2,4-Dinitrophenylhydrazine reacts with carbonyl compounds (aldehydes/ketones) in the presence of mineral acid to form 2,4-dinitrophenylhydrazone, which can be readily isolated and recrystallized.

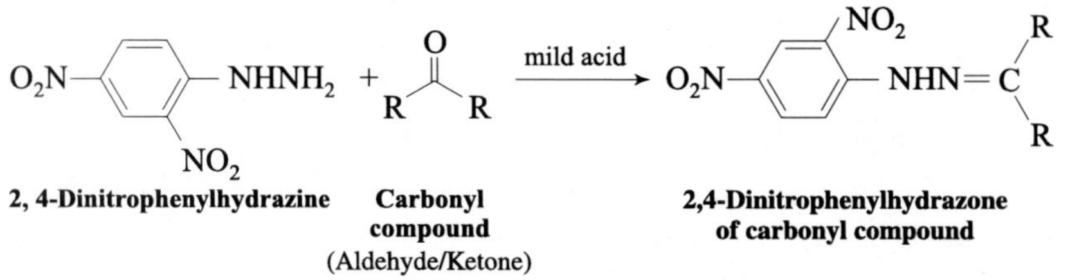

2, 4-Dinitrophenylhydrazine   Carbonyl compound (Aldehyde/Ketone)   2,4-Dinitrophenylhydrazone of carbonyl compound

Procedure: Place carbonyl compound (1.0 mL) in a boiling tube. Dissolve in minimum amount of water or alcohol (2-3 mL). Add dropwise, 2, 4-DNP reagent (5.0 mL, 2,4-DNP reagent in water for water soluble carbonyl and 2,4-DNP reagent in alcohol for water insoluble carbonyl compound). Allow the solution to stand at room temperature. Filter the precipitated solid, wash first with very dilute hydrochloric acid and then with cold water. In case no solid separates on keeping at room temperature, cool the reaction mixture in ice-water bath and scratch the inner side walls of the boiling tube to induce precipitation and then proceed as discussed above. (In certain cases, heating the solution on water bath for 10-15 minutes may be required). Dry the precipitated solid and recrystallize a part of crude sample using ethanol. Filter the crystallized solid, dry on a porous clay plate and record its melting point.

Note

- Acid present in the 2,4-DNP reagent is sufficient to catalyze the reaction. No more acid needs to be added to the reaction mixture.
- Mineral acid in the reagent also helps to keep 2,4-dinitrophenylhydrazine in solution.
- See Volume 1, *section 6.12.1.4* for details on theory.
- Washing the precipitate with dilute hydrochloric acid removes unreacted 2,4-DNP reagent.
- Never use 2,4-DNP in alcohol for aqueous solution of the compound, as the reagent (2,4-DNP) not the 2,4-dinitrophenylhydrazone will get

precipitated. This reagent is however soluble in dilute sulphuric acid whereas 2,4-dinitrophenylhydrazone is not.

- Aliphatic compounds produce yellow coloured product whereas aromatic compounds produce deep red coloured precipitate

Safe Waste Disposal

- Pour the filtrate into organic flammable waste container.

2. Semicarbazone: Semicarbazones of aldehydes and ketones can be prepared by the reaction of aldehydes/ketones with semicarbazide hydrochloride in presence of sodium acetate.

$$H_2N-NH-\overset{\overset{\textstyle O}{\|}}{C}-NH_2.HCl + CH_3COONa \rightarrow H_2N-NH-\overset{\overset{\textstyle O}{\|}}{C}-NH_2 + CH_3COOH + NaCl$$

Semicarbazide hydrochloride   **Sodium acetate**

$$RCOR + H_2N-NH-\overset{\overset{\textstyle O}{\|}}{C}-NH_2 \xrightarrow[-H_2O]{} \underset{\textstyle R}{R-C=N-NHCONH_2}$$

Carbonyl compound   **Semicarbazide**
(Aldehyde/ Ketone)

Semicarbazone of carbonyl compound

Procedure: Dissolve semicarbazide hydrochloride (1.0 g) and anhydrous sodium acetate (1.5 g) in water (5-6 mL) taken in a boiling tube. Warm the contents on water bath, if required, to get a clear solution. Then, add aldehyde/ketone (1.0 mL). Allow the mixture to stand for 10-15 minutes at room temperature. Scratch the side walls with glass rod. If no precipitate is formed in cold, warm the contents on a water bath at 60°C for 10-15 minutes. Cool the solution in ice-bath and filter the precipitated solid, wash with cold water. Dry the solid and recrystallize a part of the crude sample using hot water or aqueous ethanol. Filter the crystallized solid, dry on a porous clay plate and record its melting point.

Note

- See Volume I, *section 6.12.1.1,* semicarbazone formation for details on theory.

- Acetic acid liberated during the reaction will catalyse the formation of semicarbazone. Therefore, sodium acetate should not be added in excess.

- Prepare fresh solution of semicarbazide, as prolonged keeping of neutralized solution will lead to aerial oxidation of the semicarbazide.

- Avoid prolonged heating with semicarbazide as it forms biurea ($NH_2CONHNHCONH_2$, melting point 258°C). Biurea is soluble in water and semicarbazones are not.

Safe Waste Disposal

- Dilute the filtrate with water and add 5% aqueous HCl dropwise until pH ~5 (slightly acidic). Run down the sink.
- Pour the filtrate from recrystallization in flammable organic waste container.

3. Oxime: Oximes of carbonyl compounds can be prepared by the reaction of aldehydes/ketones with hydroxylamine hydrochloride in the presence of sodium acetate.

$$H_2N{-}OH.HCl \quad + \quad CH_3COONa \quad \longrightarrow \quad H_2N{-}OH \quad + \quad CH_3COOH + NaCl$$

Hydroxylamine hydrochloride   **Sodium acetate**   **Hydroxylamine**

$$RCOR \quad + \quad H_2N{-}OH \quad \xrightarrow{-H_2O} \quad \underset{\underset{R}{|}}{R{-}C}{=}N{-}OH$$

Carbonyl compound   **Hydroxylamine**   **Oxime of carbonyl compound**
(Aldehyde/ Ketone)

Procedure: Dissolve hydroxylamine hydrochloride (1.0 g) and anhydrous sodium acetate (1.5 g) in water (5-6 mL) taken in a boiling tube. Warm the contents on water bath, if required, to get a clear solution.) Add a solution of aldehyde/ ketone (1.0 mL) in ethanol/water (2-3 mL) to it. Cool the solution in ice bath. Scratch the inner sides of the boiling tube with the help of a glass rod to induce solid formation, if required. If no solid precipitates, heat the mixture for 10-15 minutes on boiling water bath. Cool and filter the precipitated solid, wash with cold water. Dry and recrystallize a part of crude sample using dilute alcohol. Filter the crystallized solid, dry on a porous clay plate and record its melting point.

Note

- See Volume I *section 6.12.1.3*, Addition-Elimination reactions of carbonyl compounds, for detailed theory.
- Liberated acetic acid will catalyze the reaction.
- Reaction with hydroxylamine is reversible. Avoid contact with strong acids as it may hydrolyze hydroxylamine.
- In case reaction mixture is turbid after mixing the organic compound, add ethanol (2-3 mL).
- In case ethanol is added, dilute the reaction mixture with water (4-5 mL), after heating on water bath, cool the contents in ice-water bath and filter the precipitated solid.

Safe Waste Disposal

- Place all the filtrates in aqueous waste container.

4. Iodoform: α-Methylketones/acetaldehyde/alcohols having CH_3-CHOH-group are oxidised with iodine in the presence of a base like sodium hydroxide. The hydrogen atoms on the α-carbon are first substituted by iodine to form triiodomethyl carbonyl compound. This triiodo compound further reacts with the base (OH^- ions) to produce iodoform and carboxylic acid salt.

$$\begin{array}{c} H_3C \\ \diagdown \\ \quad\quad C=O \ + \ 4NaOH \ + \ 3I_2 \ \xrightarrow{\ KI\ } \ CHI_3\downarrow \ + \ RCOONa + 3NaI + 3H_2O \\ \diagup \\ R \end{array}$$

α-Methyl ketone/
Acetaldehyde

Iodoform
(Yellow ppt.)

Procedure: Place carbonyl compound (1.0 mL), water (10.0 mL) and 10% aqueous sodium hydroxide (5.0 mL) in a conical flask. To this mixture add iodine solution [(4.2 g iodine and potassium iodide (8.3 g) dissolved in water (30.0 mL)] dropwise with constant shaking till the colour of iodine persists. Heat the contents on water bath at 60°C till yellow precipitate of iodoform settles down. If the colour of iodine still persists in the solution, add 10% aqueous sodium hydroxide solution dropwise till the colour of iodine gets discharged. Cool the hot solution, filter, wash well with cold water, dry and recrystallize a fraction of the crude sample from alcohol. Report the melting point of the crystallized sample.

Safe Waste Disposal
- Place the test solution in a beaker and dropwise add acetone to consume any unreacted iodine. Filter iodoform and transfer it to the halogenated waste container. Neutralize the filtrate to pH ~7 with 5% aqueous hydrochloric acid and run down the sink.

5. Carboxylic acids by oxidation of aldehydes: Aldehydes like benzaldehyde are oxidised to the corresponding acids by using alkaline potassium permanganate. For detailed theory, see Volume 1, *section 6.11.3*, oxidation-reduction reaction.

$$CHO \quad \xrightarrow[\text{2. Conc.HCl}]{\text{1. Alk.KMnO}_4} \quad COOH$$

Benzaldehyde

Benzoic acid

Procedure: Take a mixture of benzaldehyde (1.0 mL), sodium carbonate (1.0 g) and water (20.0 mL) in a round bottomed flask fitted with a reflux condenser. Add a saturated solution of potassium permanganate (1.5 g) taken in minimum amount of water (10-15 mL) to the round bottomed flask. Add 1-2 pumice stones and reflux the contents on wire gauze for 30-45 minutes or till the oily droplets persist. Allow the contents to cool down to room temperature and filter the solution to remove the precipitated manganese dioxide. Transfer the filtrate to 250 mL beaker. Acidify the contents with concentrated hydrochloric acid and then add solid sodium sulphite till the brown precipitate of any residual manganese dioxide completely dissolves and the supernatant liquid becomes colourless. Allow the solution to cool to room temperature and filter the separated colourless crystals of benzoic acid. Wash well with minimum amount of cold water, dry and recrystallize a small part of the crude sample with hot water. Report the melting point of the crystallized sample.

Note

- Avoid adding excess of water during workup as some carboxylic acids are soluble in water even at room temperature.
- Avoid inhaling sulphur dioxide as it is injurious for respiratory tract.
- Potassium permanganate is a skin irritant. Avoid direct contact with skin and eyes. See MSDS before use (https://pubchem.ncbi.nlm.nih.gov)
- This method can be used for oxidation of other aromatic aldehydes also. However, aromatic aldehydes with electron donor groups like NH_2, OH cannot be oxidized by this method as highly activated ring gets oxidized preferentially.

Safe Waste Disposal

- Transfer all the aqueous filtrates to hazardous waste container.
- Solvent from recrystallization should be discarded to organic waste container.

Post Lab Questions

1. Compare the reactivities of the following towards nucleophilic addition-elimination reaction:
   (a) CH_3CHO and CH_3COCH_3
   (b) CH_3CHO and CH_3CH_2CHO
   (c) HCHO and other aliphatic aldehydes.

2. How will you distinguish between the following in the lab:
   (a) Benzaldehyde and acetophenone
   (b) Acetone and acetaldehyde
   (c) Acetaldehyde and formaldehyde
   (d) Benzaldehyde and formaldehyde
   (e) 2-Pentanone and 3-pentanone

3. 2,4-Dinitrophenylhydrazone of aliphatic aldehydes is light yellow whereas for aromatic aldehydes is deep orange. Explain.

4. Explain with reasons:

   (a) Benzalehyde gives Tollens' test but does not give Fehling's test.

   (b) Acetaldehyde gives instant pink colour with Schiff's reagent while benzaldehyde reacts in 5 minutes.

   (c) Formic acid gives Tollens' test.

5. What is Tollens' reagent and why it should be freshly prepared before use?

6. Acetylacetone does not give iodoform test. Explain.

7. Which of the following give iodoform test and why:

   (a) 1-Phenylethanol

   (b) Ethanol

   (c) Ethylacetoacetate

   (d) Acetophenone

   (e) Acetic acid

   (f) Isopropanol

   (g) Iodoacetone

8. During Tollens' test, why is it necessary to decompose metallic silver immediately after formation. Explain.

9. What is the role of Rochelle's salt in the Fehling's solution test?

10. Nucleophilic addition-elimination of aldehydes/ketones with ammonia derivatives like 2,4-dinitophenylhydrazine, hydroxylamine hydrochloride and semicarbazide are carried out at controlled pH. Explain.

11. Can you distinguish between formaldehyde and formic acid using Tollens' test or Fehling's test. Give reasons.

12. Give the oxidation product of the following with alkaline potassium permanganate:

    (a) p-Xylene

    (b) t-butyl benzene

13. Distinguish between benzaldehyde and cinnamaldehyde using:

    (a) Preliminary test

    (b) UV spectroscopy

    (c) IR spectroscopy

    (c) Chemical test

14. Distinguish between o-hydroxybenzaldehyde and p-hydroxybenzaldehyde using:

    (a) Physical test

    (b) IR spectroscopy

15. Distinguish between acetaldehyde and acetone using:

    (a) NMR spectroscopy

    (b) IR spectroscopy

4.3.2.2.2 *Esters*

Esters are derivatives of carboxylic acids where the –OH of carboxyl group is replaced by –OR, R being an alkyl or aryl group. They are normally colourless liquids but some of them like phenyl benzoate (melting point 71°C), phenyl salicylate or "salol" (melting point 43°C) are crystalline solids at room temperature. Most of the esters are sparingly soluble in water however, the methyl or ethyl esters of formic acid, oxalic acid, tartaric acid, succinic acid, etc., are water soluble. Esters are characterised by fruity smell and find applications in synthetic flavours, perfumes, cosmetics, etc. Naturally occurring fats and oils are triesters of glycerol with fatty acids (C_{12}-C_{20}). Table 4.14 lists the classification test for esters.

Table 4.14   Classification Tests for Esters

S. No.	Classification test	Test category	Observation	Inference
1.	Odour	Preliminary test	Fruity smell	Could be an ester/ alcohol
2.	Hydrolysis Test	Group test	Pink colour of phenolphthalein fades	Ester
3.	Hydroxamic acid test*	Special test	Burgandy or magenta colour	Ester

*This test is also given by amines, anhydrides, amides and acid chlorides. It should be taken as +ve for esters if the test for nitrogen is negative. (Note: Anhydrides and acid chlorides are normally not given for analysis at the undergraduate level).

Classification tests for esters

1. Odour: Esters possess characteristic fruity odours. Some esters like methyl and ethyl oxalate/phthalates are odourless. It is therefore suggested to get aquainted with the odours for common esters before arriving at a conclusion. Also, odour should only be taken as a preliminary guide to proceed further. Table 4.15 gives the characteristic odours of some esters.

Table 4.15   Characteristic Odours for Some Esters

S.No.	Name	Odour
1.	Methyl/ethyl acetate	Apples
2.	Methyl salicylate	Oil of winter Green
3.	Benzyl acetate	Pear
4.	Methylphenyl acetate	Honey

2. Hydrolysis test: Esters are best hydrolysed by alkali to sodium salt of carboxylic acid and a molecule of alcohol or phenol. The reaction is called **saponification** due to the detergent action of the long chain sodium/potassium carboxylates. If a few drops of phenolphthalein are added during the hydrolysis reaction and dilute sodium hydroxide solution of right concentration is taken, the carboxylic acid formed by ester hydrolysis discharges the pink colour of phenolphthalein, due to drop in pH.

$$RCOOR' + NaOH \longrightarrow RCOONa + R'OH$$

$$\text{Ester} \qquad\qquad\qquad \textbf{Sodium carboxylate}$$

Procedure: Take dilute sodium hydroxide solution (1%, 2.0 mL) in clean test tube and add phenolphthalein (1-2 drops). The colour changes to pink. Now add the given organic compound (0.1 mL). If the pink colour is discharged in cold, add more of dilute sodium hydroxide solution till the permanent light pink colour persists. Put a glass rod and heat the contents on a small Bunsen burner flame for 1-2 minutes. Discharge of the pink colour of phenolphthalein on heating confirms the presence of ester group.

Note

- For details on theory, see Volume 1, *section 6.10.3*, hydrolysis of esters.
- This test will also be given by other acid derivatives which give carboxylic acid on hydrolysis. For example, anhydrides, amides, lactones, etc.
- Perform blank test to rule out any discrepancy.
- The carboxylic acid formed by hydrolysis neutralizes the alkali and drops the pH. This leads to discharge of pink colour. It is very important therefore that sodium hydroxide solution is very dilute and is not added in excess.

Safe Waste Disposal

- Extract the aqueous solution with ether (2 × 15 mL) to remove alcohol. Transfer ethereal layer to flammable organic waste container.
- Neutralize the aqueous layer to pH ~7 with 5% aqueous hydrochloric acid before placing it in the container for aqueous inorganic waste or running it down the drain.

3. Hydroxamic acid test: Esters react with hydroxylamine in the basic medium to form hydroxamic acid and an alcohol. The solution is acidified and then heated in the presence of aqueous ferric chloride solution. A burgundy or magenta colour is obtained due to formation of ferric hydroxamate complex.

$$RCOOR' + NH_2OH \longrightarrow RCONHOH + R'OH$$

$$\text{Ester}$$

$$3RCONHOH + FeCl_3 \longrightarrow \left[R\overset{O}{\underset{HN-O}{\diagdown\!\diagup}} \right]_3 Fe + 3HCl$$

Ferric hydroxamate complex
(Magenta or burgundy coloured complex)

Note

- Some phenols react with ferric chloride to give similar coloured products. However, the acidic solution used here usually converts a ferric phenolate complex into a free phenol, which is not deeply coloured.

- This reaction is typical for compounds bearing acyl group. Therefore, many amides, anhydrides, acid chlorides also give positive test. However, they can be confirmed by other chemical tests.

Procedure: Perform the following preliminary test before doing the hydroxamic acid test.

Preliminary test: Dissolve the given compound (1-2 drops or 0.05 g) in 95% ethanol (1.0 mL) and add hydrochloric acid (1.0 mL, 1M). Now add dropwise, aqueous ferric chloride solution (5%). Note the colour produced when one drop ferric chloride is added to the solution. If an orange, red, blue, or violet colour is produced, the following test for the acyl group is not suitable for the given compound. For others proceed as follows:

Hydroxamic acid test: Take the given compound (1-2 drops) in a clean test tube. Add hydroxylamine hydrochloride (0.2 g) and NaOH solution (5.0 mL, 10%). Gently boil the mixture on Bunsen burner for 1-2 minutes. Cool and acidify with dilute HCl, cool again and then add a few drops of $FeCl_3$ solution until the burgundy or magenta colour of ferric hydroxamate complex persists. Compare the colour with that produced in the preliminary test above.

Note

- Compounds which give orange, red, blue, or violet colour with ferric chloride cannot be identified by this reaction. For example, methyl and ethyl salicylate give violet colour with ferric chloride.

- Colour produced from hydroxamic acid test should be different from that in the initial preliminary test with ferric chloride.

- Hydrochloric acid is added before adding ferric chloride solution to neutralize the reaction medium and rule out any discrepancy which may arise due to precipitation of ferric hydroxide in the alkaline medium.

Safe Waste Disposal

- Place the test solution in the container for inorganic waste.

4. Hydroxamic Acid for esters (Grove tile Test):   Add 1-2 drops each of the following compounds to a groove tile, in the order mentioned below:

- The given compound
- Saturated solution of sodium hydroxide
- Concentrated solution of hydroxylamine hydrochloride.

Keep the solution in the groove tile for 2-3 minutes then add the following:

- Concentrated hydrochloric acid to acidify.
- Now add ferric chloride solution.

Development of magenta or burgundy colour confirms the presence of ester.

5. Spectroscopic identification of esters

Table 4.16 Spectral Data of Esters

S. No.	Type of spectrum	Peak value	Peak attribution	Remarks
1.	IR	$1735\text{-}1750 \ cm^{-1}$	$>C=O$ str.	Strong
		$1250 \ cm^{-1}$	C-O str.	Singlet or doublet
2.	NMR	$3.5\text{-}4.8 \ \delta \ *$	$>COOC(H)<$	-
		$2.1\text{-}2.5 \ \delta$	$>C\underline{H}COOR$	-

Derivatives of ester

1. Saponification of Ester: *[This approach can be used in case the alcohol is water insoluble and acid part is water soluble/insoluble].* Base promoted hydrolysis of an ester, also called saponification, is a *Nucleophilic acyl substitution reaction,* where the corresponding sodium salt of carboxylic acid and alcohol is obtained. Acidification of the sodium salt of carboxylic acid gives corresponding carboxylic acid (For detailed theory, see Volume 1, *section 6.10.3*). For example, esters of benzoic acid and salicyclic acid give respectively benzoic acid and salicylic acid.

$$\underset{\textbf{Ester}}{RCOOR^1} + NaOH \longrightarrow RCOONa + R^1OH \xrightarrow{\text{HCl}} \underset{\textbf{Carboxylic acid}}{RCOOH} + R^1OH + NaCl$$

The alcohol part can be identified by distillation of the reaction mixture after hydrolysis is complete.

Note

- Acid hydrolysis of esters is reversible, therefore avoided.

Procedure: Place the given ester (1.0 mL), 10% aqueous sodium hydroxide solution (15.0 mL) and a few pumice stones in a 50 mL round bottomed flask fitted with a water condenser. Reflux the mixture over wire gauze gently for 25-30 minutes or till reaction is complete (fruity smell will disappear). The following observations can be made on completion of the reaction:

(a) An oily layer on top of aqueous alkaline solution. This is due to insoluble alcohol part (e.g., cyclohexanol, benzyl alcohol, etc.). If observed, distill to about half of the reaction mixture and saturate the distillate with potassium carbonate. Two layers will be formed due to alcohol and aqueous phase. Separate the alcohol part using a dropper, dry over anhydrous sodium sulphate and identify the alcohol using boiling point measurement.

Next, acidify a fraction of the residual solution in the R.B. flask with dilute sulphuric acid. If insoluble precipitate (due to water insoluble carboxylic acid) is formed, proceed as in (b) otherwise proceed as in (c).

(b) Acidify the entire residual solution with dilute sulphuric acid, filter the insoluble acid, recrystallize with water or aqueous alcohol and identify the acid part using melting point determination.

(c) In case no solid separates on acidification, take the residual solution (5.0 mL) in R.B. flask (50 mL), concentrate to half (~2.0 mL), add a drop or two of phenolphthalein (ensure permanent pink colour) and add just enough dilute HCl to get light pink colour and prepare the SBT derivative with the resulting solution. (For procedure, see derivatives of carboxylic acids, *section 4.3.2.1.1*, page 130.

Note

- Water soluble alcohol cannot be identified by this method.

2. Green saponification of esters: Saponification of esters may also be carried out by refluxing the reaction mixture (See saponification of esters, page 167) over microwave oven at ~90-100°C. The hydrolysis proceeds to completion in 5-10 minutes. The workup may be attempted as in (a)/(b)/(c) above to identify the alcohol and acid part.

Note

- Green components of the reaction are:
    - (a) Energy efficient process, since microwave is used as heating source.
    - (b) Faster reaction

3. 3,5-Dinitrobenzoate of the alcohol part: [*To be used if alcohol is water soluble and cannot be identified by saponification method*].

The ester is subjected to trans-esterification reaction with 3,5-dinitrobenzoic acid in the presence of traces of concentrated sulphuric acid, as catalyst, to identify the alcohol or phenol part directly.

$$RCOOR' + \text{3,5-Dinitrobenzoic acid} \xrightarrow{H_2SO_4} \text{3,5-Dinitrobenzoate} + RCOOH$$

Ester   3,5-Dinitrobenzoic acid   3,5-Dinitrobenzoate

Procedure: Take the given ester (1.0 mL) in a clean and dry boiling tube, add 3,5-dinitrobenzoic acid (1.25 g) and concentrated sulphuric acid (5-6 drops). Stir the mixture and heat on a low flame on sand bath, continue occasional stirring while heating. The solution becomes dark coloured on heating but mild heating prevents charring of the ester. After ~10-15 minutes, stop heating and check for completion of reaction by removing a drop of the reaction mixture and adding it over small amount of ice taken in a beaker. Separation of solid confirms completion of reaction. Else, continue heating until the reaction is complete. Now, pour the entire reaction mixture over ice (~50-100 g) taken in a 100 mL beaker and stir. Filter the solid and transfer it to a fresh 100 mL beaker. Treat it with saturated sodium bicarbonate solution until the effervescence cease. Filter the remaining solid, recrystallize from alcohol and determine its melting point to identify the alcohol part of the ester.

Note

- The absence of a pleasant smell characteristic of ester indicates completion of reaction. In case of water soluble alcohols, disappearance of oily layer indicates completion of reaction.
- In case of water insoluble alcohol, progress of the saponification reaction can be monitored on TLC.
- 15% Alcoholic KOH can also be used for saponification of water insoluble esters. Alcohol part in such cases however is identified by spectral data and boiling point of esters. Iodoform test can not be used in such cases.
- Esters with boiling point < 110°C saponify very quickly whereas esters with higher boiling points take longer time (1-2 h).
- In case the alcohol part is water soluble like ethanol or isopropanol, no oil separates on completion of saponification. In such cases, prepare the iodoform from a part of the distillate to confirm the alcohol else proceed as in **method 3**, to directly identify the alcohol part.
- During the preparation of 3,5-dinitrobenzoates, washing with aqueous sodium bicarbonate removes any unreacted 3,5-dinitrobenzoic acid.

Safe Waste Disposal

- Place all the test solutions in aqueous solution container after neutralizing the filtrates with 5% aqueous NaOH or 5% aqueous HCl as required.

4. Green trans-esterification reaction for identification of soluble alcohol part: In a clean and dry round bottomed flask, take a mixture of 3,5-dinitrobenzoic acid (1.25 g), ester (1.0 mL) and concentrated sulphuric acid (5-6 drops). Reflux with stirring at 70°C on a microwave oven for ~7 minutes. Work up as explained in **method 3**, above.

Advantages of the green process:
- Less energy is consumed
- Time required reduced considerably.

Post Lab Questions

1. Why concentrated alkali cannot be used in the hydrolysis test for esters?

2. A student gets an unknown phenol however he misses the ferric chloride test. Do you think he will get colouration in the hydroxamic acid test? If not, why?

3. A neutral compound gives decolourization with phenolphthalein in the alkaline hydrolysis test and gives a positive hydroxamic acid test. Can you say with certainty that the compound is an ester?

4. What is the need to acidify the reaction mixture during hydroxamic acid test?

5. What is the role of alkali in the hydroxamic acid test?

4.3.2.2.3 *Alcohols*

If the unknown organic compound is neutral, does not give burning sugar smell and any characteristic tests for aldehydes, ketones and esters, they can either be alcohols or ethers.

Alcohols are compounds having general formula ROH, where R is an alkyl or aryl-alkyl group. Depending upon the nature of carbon to which the OH group is attached, the alcohols may be classified as primary, secondary or tertiary alcohols. They are mostly colourless liquids, lower members are volatile, completely miscible with water (exception: benzyl alcohol and cyclohexanol which are sparingly soluble. Most alcohols have a faint odour but glycol and glycerol are odourless. Odour becomes less pronounced as the number of carbons increase. Commercially, alcohols are found in beverages, and are useful fuels to blend with petrol. Ethanol is used as a solvent in many organic reactions, in toiletries and for sterilizing surgical instruments.

Chemically, alcohols are neutral however, when reacted with alkali metals like sodium or potassium, they liberate hydrogen gas showing weak acid character (especially primary alcohols). Being weak acids, they do not liberate carbon dioxide with sodium bicarbonate. The acidity of alcohols progressively decreases as we move from primary through secondary to tertiary alcohols. Structural effects on acidity are complex. However, it is well known that liquid samples of alcohols follow the following order of acidity:

$$CH_3OH > CH_3CH_2OH > (CH_3)_2CHOH > (CH_3)_3COH$$

The order can be explained on the basis of the stability of the conjugate base. It is presumed that bulky groups about the oxygen atom destabilize solvation of alkoxide ion, thus decreasing its stability. Table 4.17 lists the classification tests for alcohols.

Table 4.17   Classification Tests for Alcohols

S. No.	Classification test	Test category	Observation	Inference
1.	Sodium metal test[a]	Group test	Brisk effervescence	Alcohol
2.	Ceric ammonium nitrate test	Special test	Blood red colouration[b]	Alcohol
3.	Iodoform test	Special test	Yellow precipitate	CH_3CHOH- group
4.	Lucas test	Special test	Cloudiness at different rate	1°/2°/3° alcohol
5.	Oxidation with chromic acid	Special test	Green or blue colouration	Primary or secondary alcohols[c].

a: This test can be performed after ruling out carboxylic acids, phenols, primary and secondary amines and methyl ketones; **b:** red colour fades on standing due to reduction of Ce(IV) to Ce(III). **c:** distinguishes primary and secondary alcohols from tertiary alcohols.

Classification tests for alcohols

1. Sodium metal test: Many alcohols react with sodium metal to liberate hydrogen gas and form the sodium alkoxide. The rate of reaction depends on the structure of the alcohol.

$$2ROH + 2Na \longrightarrow 2RONa + H_2$$

$\text{Alcohol} \quad \text{Sodium} \qquad \text{Sodium Alkoxide}$

Procedure: Take the given organic compound (1.0 mL) in a clean and dry test tube and add some anhydrous calcium sulphate or sodium sulphate. Allow the contents to stand for 5 minutes. Decant the supernatant liquid into another dry test tube. Add a small piece of freshly cut sodium into the test tube. Sodium metal dissolves with evolution of brisk effervescence. This confirms the presence of alcohol.

Note

- Presence of traces of moisture causes the sodium metal test to be positive. Therefore, drying the sample with anhydrous calcium sulphate or sodium sulphate is important before adding a piece of sodium metal.

- Compounds containing active hydrogen like phenols, thiols, primary or secondary amines, methyl ketones, carboxylic acids and terminal alkynes also react with sodium to liberate hydrogen gas. These groups however can be detected by other distinguishing tests

- This test is most useful with alcohols containing three to eight carbon atoms. Lower alcohols are difficult to obtain in anhydrous condition and evolution of hydrogen gas is very slow in case of higher alcohols.

Safe Waste Disposal

- Add ethanol dropwise, with shaking, to the test solution until all the sodium dissolves to form alkoxide. Allow the contents to stand for one hour. Dilute the reaction mixture with water (10.0 mL), neutralize with aqueous hydrochloric acid (5%) and transfer to the aqueous waste solution container.

2. Ceric ammonium nitrate test: The ceric ammonium nitrate reagent (CAN) reacts with small to medium chain alcohols (having up to 10 carbons) to form red coloured alkoxy cerium (IV) complex.

$$(NH_4)_2[Ce(NO_3)_6] + ROH \longrightarrow (NH_4)_2[Ce(OR)(NO_3)_5] + HNO_3$$

Ceric ammonium   **Alcohol**   **Red**
nitrate (Yellow)

Procedure: Take ceric ammonium nitrate solution (1.0 mL) in a test tube and add the unknown compound (3-4 drops or 0.1 g) taken in 1-2 mL water. Shake thoroughly and note if the colour of the yellow coloured reagent changes to red. Appearance of red colour confirms presence of alcohol.

Alternatively, take dilute nitric acid (1-2 mL) and add 2-3 crystals of solid ceric ammonium nitrate, shake to dissolve the crytals. Add the given organic compound (2-3 drops). Appearance of red colour confirms an alcohol.

Note

- Use freshly prepared CAN solution only.
- Glycols, polyhydroxy compounds including carbohydrates, hydroxy aldehydes, hydroxy ketones, and hydroxy acids also give red solutions. Phenols, however, give brown colouration or precipitate.
- Water insoluble compounds may be dissolved in dioxan before adding to CAN solution. A blank test should be performed as dioxan sample is sometimes contaminated with alcohol.
- An immediate red colouration on mixing should only be taken as positive test.
- Red colouration fades on standing due to reduction of Ce(IV) to Ce(III).

Safe Wase Disposal

- Place the test solution in the aqueous waste solution container.

3. Iodoform test: See the classification test of aldehydes and ketones, *section 4.3.2.2.1,* page 155.

4. Lucas test: This reaction is used to distinguish between primary, secondary and tertiary alcohols. Lucas reagent is a solution of anhydrous zinc chloride in concentrated hydrochloric acid. The reaction of alcohols with Lucas reagent proceeds *via* S_N1 mechanism. The rates of reaction of primary, secondary and tertiary alcohols with hydrochloric acid are different due to difference in their ability to form stable carbocations, which react further with chloride ion to form alkyl chlorides. The tertiary carbocation, being most stable, is formed much faster than secondary carbocations, which in turn are formed faster than primary carbocations. Since chloride ion is a weak nucleophile and hydroxide ion is a poor leaving group, the reaction cannot possibly proceed *via* S_N2 mechanism. The reactivity of alcohols with the Lucas reagent will therefore follow the order:

Tertiary > Secondary > Primary

In fact, no reaction occurs in case of primary alcohols at room temperature.

It should be noted however, that any other alcohol which is capable of forming carbocation easily (For example, allyl alcohol and benzyl alcohol), also reacts readily with Lucas reagent. The visual change in the reaction is based on the fact that alcohols are soluble in Lucas reagent but the alkyl chlorides are not. Hence, the formation of latter is seen as cloudiness in the reaction mixture. The presence of zinc chloride, a good Lewis acid, makes the reaction mixture even more acidic than HCl alone and enhances the formation of carbocations by increasing the leaving group ability of the hydroxide ion.

(a) Increased acidity of HCl is due to formation of coordinate complex:

$$\overset{\delta+}{H}\text{-------}\overset{\delta-}{Zn}Cl_3$$

(b) Increased rate of formation of carbocation is due to formation of Lewis acid-base complex with $ZnCl_2$:

$$ROH + ZnCl_2 \rightleftharpoons R-\underset{H}{\overset{\overset{\delta+}{O}}{|}}-\overset{\delta-}{Zn}Cl_2 \xrightarrow{HCl} RCl + [Zn(OH)Cl_2]^{\ominus}$$

Lewis acid base complex

$$[Zn(OH)Cl_2]^{\ominus} + H_3O^{+} \rightleftharpoons ZnCl_2 + 2H_2O$$

$$R-CH_2-OH \xrightarrow{(ZnCl_2 + HCl)} \text{No reaction}$$
1° Alcohol

$$R-\underset{R}{\underset{|}{CH}}-OH \xrightarrow{(ZnCl_2 + HCl)} R-\underset{R}{\underset{|}{CH}}-Cl$$
2° Alcohol   **White turbidity within five minutes**

$$R-\underset{R}{\overset{R}{\underset{|}{\overset{|}{C}}}}-OH \xrightarrow{(ZnCl_2 + HCl)} R-\underset{R}{\overset{R}{\underset{|}{\overset{|}{C}}}}-Cl$$
3° Alcohol   **White turbidity instantly**

Procedure: Take the given compound (0.1mL) in a clean and dry test tube and add Lucas reagent (1.0 mL). Shake the contents vigorously to ensure thorough mixing. Note the time taken for cloudiness to appear. Instant cloudiness indicates tertiary alcohol whereas secondary alcohols give cloudiness in 5-10 minutes. Primary alcohols do not react at room temperature.

Note

- This test should be performed in fume hood only.
- This test is suitable for alcohols having less than six carbons (water soluble) because the solubility of the alcohol in the Lucas reagent forms the basis of this test.
- Carbohydrates, which are polyhydroxy aldehyde or ketones, also give a positive Lucas test.
- Other primary alcohols capable of forming a stable carbocation also give positive Lucas test. For example, benzyl and allyl alcohol.

Safe Waste Disposal

- Transfer the test solution to water (20.0 mL) in a beaker. Neutralize the solution by carefully adding solid sodium carbonate until basic to pH paper (caution: foaming). Filter the residue of zinc hydroxide by vacuum filtration. Transfer the residue of zinc hydroxide to inorganic solid waste container. Neutralize the filtrate to pH ~7 with 5% aqueous hydrochloric acid. Now, pour the filtrate down the drain or pour it into aqueous inorganic waste container.

5. Oxidation with chromic acid (Jones reagent): This test can be used for distinguishing tertiary alcohols from primary and secondary alcohols. Primary and secondary alcohols, due to the presence of replaceable hydrogen, are oxidised by Jones reagent, which is reduced to green or blue coloured chromic sulphate. Chromium is reduced from Cr(VI) to Cr(III), during the reaction. The reaction is given by all primary and secondary alcohols, irrespective of the chain length or molecular weight.

$$3RCH_2OH + 4CrO_3 + 6H_2SO_4 \longrightarrow 3R\overset{\displaystyle O}{\overset{\|}{-C}}-OH + 9H_2O + 2Cr_2(SO_4)_3$$

1° Alcohol   (Orange-red)   **Carboxylic acid**   Blue to green

$$3R_2CHOH + 2CrO_3 + 3H_2SO_4 \longrightarrow 3R\overset{\displaystyle O}{\overset{\|}{-C}}-R + 6H_2O + Cr_2(SO_4)_3$$

2° Alcohol   (Orange-red)   **Ketone**   Blue to green

$$3R\overset{\displaystyle O}{\overset{\|}{-C}}-H + 2CrO_3 + 3H_2SO_4 \longrightarrow 3R\overset{\displaystyle O}{\overset{\|}{-C}}-OH + 3H_2O + Cr_2(SO_4)_3$$

Aldehyde   **Carboxylic acid**   Blue to green

Procedure: Take acetone (1.0 mL) in a clean, dry test tube and add the given compound (1-2 drops or 0.010 g of a solid). Add Jones reagent (1.0 drop) and immediately note the colour change. An opaque suspension with a green to blue colouration indicates primary or secondary alcohol.

Note

- Jones reagent is chromic anhydride in aqueous sulphuric acid, see Appendix II for preparation.
- A blank test with acetone should be run before arriving at a conclusion.
- Tertiary alcohols do not give any visible change within 10-15 seconds, the solution remains orange in colour. However, sulphuric acid may dehydrate tertiary alcohols to alkenes producing green Cr(III) salts. Thus, any reaction beyond 5 seconds is taken as negative.
- Aldehydes also give green colouration with Jones reagent but they can be identified by other classification tests. Ketones do not react.
- Enols may also give green colouration whereas phenols give a dark coloured solution quite different from that in the Jones test. Perform the test with ethanol and acetone to see how +ve and –ve tests come before arriving at any conclusion.

Safe Waste Disposal

- Neutralize the aqueous solution by adding solid sodium carbonate until pH is ~7. Then transfer the test solution to the hazardous waste container.

6. Spot test for alcohols

(a) *CAN test:* Spot the following compounds sequentially one over the other on a Whatman paper strip, as detailed below:

- Liquid compound or its concentrated solution, if solid (spot 2-3 times)
- Concentrated CAN reagent (spot 2-3 times)

Development of blood red spot, which fades on standing, confirms alcohol.

7. Spectroscopic identification of alcohols

Table 4.18 Spectral Data of Alcohols

S. No.	Type of spectrum	Peak value	Peak attribution	Remarks
1.	IR	3584-3650 cm^{-1} 3200-3400 cm^{-1} 1000-1260 cm^{-1}	O-<u>H</u> Str.[*](free) O-H Str.[*](H-bonded) C-O str.	Sharp, Strong[***] Broad, strong -
2.	NMR	0.5-5 δ[**] 3.2-3.8 δ	O<u>H</u> >C<u>H</u>OH	Broad, replaceable with D_2O

[*] Position and shape of the peak depends on many factors like solvent, concentration, etc.
[**] Depending on the concentration of solution.
[***] If the spectrum is run neat or in concentrated solution/solvent is polar/compound shows strong intermolecular hydrogen bonding.

Derivatives of Alcohols

1. 3,5-Dinitrobenzoate: Alcohols are known to form crystalline 3,5-dinitrobenzoates on reaction with 3,5-dinitrobenzoyl chloride.

$$ROH \;+\; \underset{\text{3,5-Dinitrobenzoyl chloride}}{\text{(O}_2\text{N, NO}_2\text{-C}_6\text{H}_3\text{-COCl)}} \longrightarrow \underset{\text{3,5-Dinitrobenzoate}}{\text{(O}_2\text{N, NO}_2\text{-C}_6\text{H}_3\text{-COOR)}} \;+\; HCl$$

Alcohol | 3,5-Dinitrobenzoyl chloride | 3,5-Dinitrobenzoate

Procedure: Place alcohol (2.0 mL) and 3,5-dinitrobenzoyl chloride (0.5 g, See Appendix II for preparation) in a dry conical flask. Heat the contents on water bath for 10-15 minutes. Cool the contents and add a few pieces of crushed ice to it. Filter the precipitated solid, wash with saturated aqueous solution of sodium bicarbonate. Recrystallize a part of the crude sample using ethanol. Filter the crystallized solid, dry on a porous clay plate and record its melting point.

Note

- Pour a drop of reaction mixture over ice to check for completion of reaction, before workup. Appearance of crystalline solid indicates completion of reaction.

- 3,5-Dinitrobenzoyl chloride is a strong irritant, avoid direct contact with the eyes and skin. Wear gloves and consult MSDS before use. (https://Pubchem.ncbi.nlm.nih.gov)

- In case of water insoluble compounds, use a slight excess of 3,5-dinitrobenzoyl chloride. This ensures complete removal of alcohol from the reaction mixture. Any unreacted alcohol will prevent the crystalline benzoate from precipitating out.

Safe Waste Disposal

- Acidify the alkaline filtrate with 5% aqueous HCl until pH ~ 5 (mildly acidic). Collect any precipitate of 3,5-dinitrobenzoic acid. Place it in solid organic waste container. Neutralize the acidic filtrate with solid sodium carbonate until pH is ~7. Run it down the sink or transfer to aqueous solution container.

2. 4-Nitrobenzoate: The procedure used for preparation of *p*-nitrobenzoates is same as described above for 3,5-dinitrobenzoate. Here, 3,5-dinitrobenzoyl chloride is replaced by 4-nitrobenzoyl chloride.

$$\text{ROH} \quad + \quad \text{4-Nitrobenzoyl chloride} \quad \longrightarrow \quad \text{4-Nitrobenzoate} \quad + \quad \text{HCl}$$

Alcohol **4-Nitrobenzoyl chloride** **4-Nitrobenzoate**

Safe Waste Disposal

- Place the aqueous filtrate into the aqueous solution container, after neutralizing to pH ~7 with solid sodium carbonate.

3. Iodoform: Alcohols having CH$_3$CHOH- group (e.g., ethanol, propan-2-ol) form iodoform on reaction with iodine and sodium hydroxide.

$$CH_3CH_2OH \xrightarrow{I_2/NaOH} CH_3CHO \xrightarrow{I_2/NaOH} I_3CCHO \xrightarrow{NaOH} CHI_3 \downarrow + HCOONa$$

Ethanol **Iodoform** (Yellow ppt.)

Procedure: See derivatives of aldehydes and ketones, *section 4.3.2.2.1,* page 161, for the procedure.

Safe Waste Disposal

- See derivatives of aldehydes and ketones, *section 4.3.2.2.1,* page 161, for clean-up.

4. Oxidation of side chain of aromatic alcohols with alkaline potassium permanganate: Like aromatic aldehydes, aryl-alkyl acohols can also be oxidised by alkaline potatssium permanganate to derivatives of benzoic acid. Presence of at least one benzylic hydrogen is a necessary and sufficient condition for the oxidation under these conditions. For example, benzyl alcohol is oxidized under the reaction conditions but 2-phenylpropan-2-ol is not.

$$\underset{\textbf{Benzyl alcohol}}{CH_2OH} \xrightarrow[\text{2. HCl}]{\text{1. Alk.KMnO}_4} \underset{\textbf{Benzoic acid}}{COOH}$$

For detailed theory, see Volume 1, *section 6.11.2.*

Procedure: Take a mixture of benzyl alcohol (1.0 mL), sodium carbonate (1.0 g) and water (1.0 mL) in a round bottomed flask fitted with a reflux condenser. Add a saturated solution of potassium permanganate (6.0 g), taken in minimum amount of water (10-15 mL), to the round bottomed flask. Add 1-2 pumice stones and reflux the contents on wire gauze for 1-2 hours. Allow the contents to cool down to room temperature, filter the brown precipitate of manganese dioxide and transfer the filtrate to a 250 mL beaker. Acidify the contents with concentrated hydrochloric acid (*check blue litmus should turn red*). Next add solid sodium sulphite in small lots, till the brown precipitate of any residual manganese dioxide dissolves completely. Allow the solution to cool to room temperature. Filter the colourless crystals of benzoic acid, which separate out. Wash the crude sample well with minimum amount of cold water, dry and recrystallize a fraction of the crude sample with hot water. Report the melting point of the crystallized sample.

Note

- Ensure the purple tinge due to potassium permanganate remains during entire refluxing period, otherwise cool the reaction mixture and add ~10 mL of the saturated solution of potassium permanganate. Continue refluxing until the reaction completes.

- Disappearance of oily droplets is an indication that the reaction is complete. Look for absence of bitter almond smell.

Safe Waste Disposal

- Place the test solutions in the inorganic waste container.

5. Green Preparation of 3,5-dinitrobenzoate: Take the given alcohol (1.0 mL) in a clean and dry round bottomed flask (50 mL), add 3,5-dinitrobenzoic acid (1.5 g) and concentrated sulphuric acid (5-6 drops). Now, reflux the contents in a microwave maintained at 70°C for ~7 minutes. Cool and pour the entire reaction mixture over ice (~50-100 g) in a 100 mL beaker and stir. Filter the solid and transfer it to another 100 mL beaker. Treat it with saturated sodium bicarbonate solution (20-25 mL) and allow to stand at room temperature until the effervescence cease. Filter the remaining solid, wash well with minimum amount of cold water. Recrystallize a part of the crude sample from alcohol and determine its melting point.

Note

- Use of hazardous thionyl chloride and phosphorous pentachloride is avoided as acid is directly converted to ester.

- Toxic and corrosive sulphuric acid is used only in catalytic amounts.

- Method is energy efficient as microwave is the heating source.

- Reaction is faster than the conventional method.

- Reaction works well for most primary alcohols except benzyl alcohol.

- Same green procedure also works well for 4-nitrobenzoates of alcohols.

Post Lab Questions

1. Alcohols are weaker acids than phenols and carboxylic acid. Explain.
2. How can you distinguish primary, secondary, and tertiary alcohols using:
    (a) Hinsberg's test
    (b) Lucas test
3. What is the role of anhydrous zinc chloride in Lucas test?
4. Why do you need to dry alcohols with anhydrous sodium sulphate before reacting with sodium?
5. Can you use anhydrous calcium chloride for drying alcohol?
6. Predict the rate of reaction of the following alcohols with sodium metal:
    Butan-1-ol, Butan-2-ol, 2-Methylpropan-2-ol
7. Predict the wave number for the –OH stretching peaks of alcohol in the following:
    (a) Concentrated solution
    (b) Dilute solution
    Why is the signal broad in case (a)?
8. Write the structure of red coloured complex of ethanol with ceric ammonium nitrate. Why does the colour fade on standing?
9. Give one test to distinguish between the following pairs:
    (a) Methanol and ethanol
    (b) Propan-1-ol and Propan-2-ol
10. Explain the terms:
    (a) Power alcohol
    (b) Methylated spirit
11. Why copper sulphate is added to commercial samples of methylated spirit?

4.3.2.2.4 *Carbohydrates*

In case you have a compound, which is aliphatic (c.f. flame test), freely soluble in water, chars on warming with concentrated sulphuric acid and gives a smell of burning sugar, a carbohydrate is indicated.

Carbohydrates are polyhydroxy aldehydes or ketones or compounds which can be converted to these on hydrolysis. Most significant members of this family include-glucose, fructose, mannose, sucrose, maltose and galactose etc. They are colourless solids, decompose on heating and therefore have no definite melting point. They are major source of energy in the body, they are essential constituents of nucleic acids (2-deoxy ribose and ribose), provide supporting framework to plants (cellulose), serve as food for young plant (starch), are sweetening agents in syrups and confectionaries (glucose and fructose), and constitute major milk sugar (galactose). Table 4.19, enlists the classification tests for carbohydrates.

Table 4.19   Classification Tests for Carbohydrates

S. No.	Classification test	Test category	Observation	Inference
1.	Charring test*	Preliminary test	Smell of burning sugar with charring	May be a carbohydrate
2.	Molisch test	Group test	Violet ring at the junction of two layers	Carbohydrate
3.	Barfoed's reagent test	Special test	Red precipitate	Monosaccharide
4.	Seliwanoff's test	Special test	Cherry red or blue precipitate	Ketose
5.	Rapid furfural test	Special test	Instant violet colouration on boiling.	Fructose/sucrose
6.	Tollens' test	Special test	Silver mirror on the inner walls of test tube	Reducing sugar
7.	Fehling test	Special test	Brick red precipitate	Reducing sugar
8.	Benedict's solution test	Special test	Brick red precipitate	Reducing sugar

*This test will also be given by many polyhydroxy compounds. See *section 4.3.1.5*, page 98, solubility in concentrated sulphuric acid for details.

Note

- Carbohydrate being a polyhydroxy compound also gives a positive CAN test, the red colour fades on standing.
- Since carbohydrates mainly exist in the cyclic form in solution, they do not react with 2,4-DNP reagent.

Classification test for carbohydrates

1. Charring test with sulphuric acid

Proceure: On a clean metal spatula, warm the given compound (0.2 g) with concentrated sulphuric acid (1-2 drops), using a small Bunsen burner flame. Smell of burning sugar is immediately observed and the compound gets charred. As the temperature is raised, CO_2, CO and SO_2 evolve and the compound leaves a black coloured residue.

2. Molisch test: This test is based on the fact that carbohydrates (hexoses and pentoses) undergo dehydrative cyclisation with concentrated sulphuric acid to form 5-hydroxymethyl furfural (HMF) or furfural. HMF or furfural then condense with Molisch reagent (See Appendix II) to form red or violet coloured complex which can be seen as a ring at the junction of the two layers.

Conversion of fructopyranose to 5-hydroxymethylfurfural

CHO
|
(CHOH)$_4$
|
CH$_2$OH
Aldohexose

5-Hydroxymethyl furfural (HMF)

Violet colour

Procedure: Dissolve the given compound (0.02 g) in water (2.0 mL) taken in a clean test tube. Add Molisch reagent (8-10 drops) and shake. Then carefully add concentrated sulphuric acid (2.0 mL) along the side of the test tube. A deep violet or red-violet coloured ring at the junction of the two layers indicates the presence of carbohydrate.

Note

- This test will be given by nucleic acids and glycoproteins also because they contain a sugar moiety.
- Sometimes a red-violet ring is obtained which on standing changes to violet.
- Use dilute solution of α-naphthol for this test
- On mixing the two layers, a violet coloured solution is obtained.

Safe Waste Disposal

- Cautiously transfer the test solution into a beaker containing 30.0 mL of water (do not attempt the reverse). Add aqueous solution of sodium carbonate till neutral to pH paper (caution: foaming!). Add a few ice chips to cool the solution to room temperature (if required). Transfer the neutral solution to the aqueous waste container.

3. Barfoed's Test-Distinction between Monosaccharides and Disaccharides/ higher saccharides: Barfoed's reagent is ~7% solution of cupric acetate in 1% acetic acid. It is used to distinguish monosaccharides like glucose and fructose from disaccharides or higher saccharides. It is based on the fact that monosaccharides, being more reactive, can reduce cupric ions in the acidic medium to red coloured cuprous oxide, whereas disaccharides and higher saccharides are unable to do so. The reaction with monosaccharide is shown below:

$$\underset{\textbf{Aldohexose}}{\overset{\displaystyle CHO}{\underset{\displaystyle CH_2OH}{|\ (CHOH)_4\ |}}} + 2Cu(CH_3COO)_2 + 2H_2O \longrightarrow \overset{\displaystyle COOH}{\underset{\displaystyle CH_2OH}{|\ (CHOH)_4\ |}} + \underset{\textbf{red ppt.}}{Cu_2O\downarrow} + 4CH_3COOH$$

Procedure: Take the given compound (1.0 mL) and Barfoed reagent (1.0 mL) in a test tube, mix well and heat the contents on a boiling water bath for 2-3 minutes. Brick red precipitate of cuprous oxide confirms monosaccharide.

Note

- Cupric ions are weaker oxidising agents in the acidic medium than the alkaline medium.
- Disaccharides may give this test on prolonged heating, due to partial hydrolysis of the acetal linkage.
- Barfoed reagent should be freshly prepared, as it does not keep well.

Safe Waste Disposal

- Transfer the test solution to the aqueous waste container.

4. Tests for distinction between reducing and non-reducing sugars: All simple sugars capable of mutarotation exist in equilibrium with the open chain form, which contains a free aldehydic or ketonic group. All such sugars reduce weak oxidising agents like Tollens' reagent, Fehling solution or Benedict's reagent and are therefore termed as **reducing sugars.** For example, glucose, fructose, mannose and galactose. The reaction being essentially the same as for the oxidation of aldehydes by these reagents (Refer to classification tests for aldehydes and ketones, *section 4.3.2.2.1, page 151,* for details). Note that ketones normally do not reduce Tollens' reagent or Fehling solution. However, ketoses behave as reducing sugar due to formation of small amount of aldose in the weak alkaline medium (**Lobry Debruyn Van Ekenstein rearrangement**).

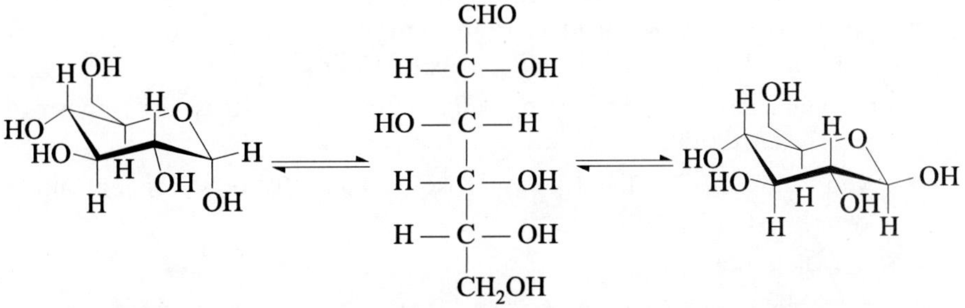

α-D(+)– Glucose
(36%)

Open chain
(Trace)

β-D(+)–Glucose
(64%)

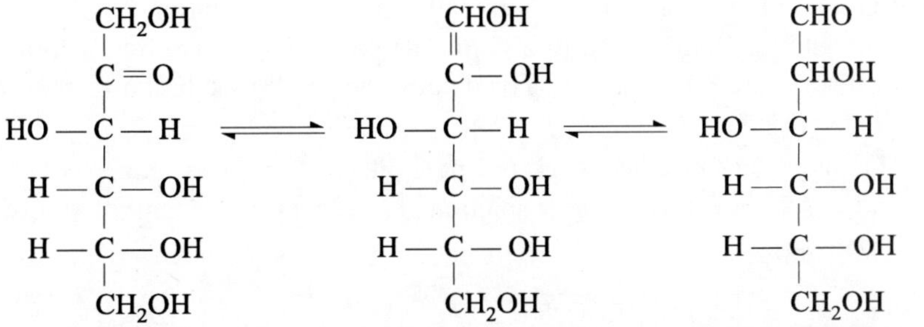

D-Fructose

Enol intermediate

D-Glucose/D-mannose

Lobry Debruyn Van Ekenstein Rearrangement

In case of monosaccharides, disaccharides or higher molecules, but not polysaccharides, existence of free OH group on anomeric carbon is necessary and sufficient condition for reducing characteristics of sugars.

(a) *Tollens' test:* Reducing sugars reduce Tollens' reagent to metallic silver and are oxidised to corresponding aldonic acid salt. The reaction is essentially the same as for aldehydes.

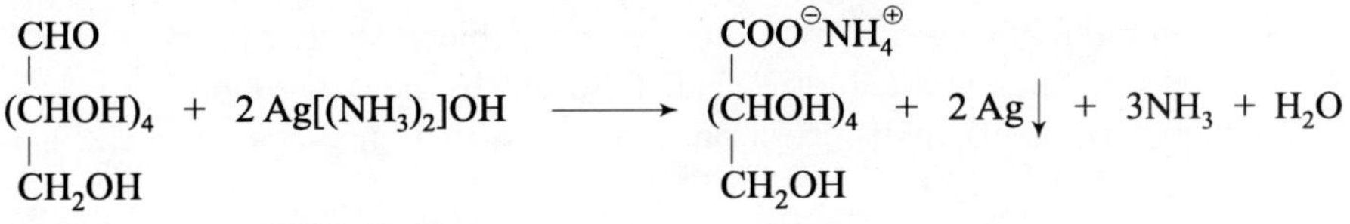

$$CHO$$
$$|$$
$$(CHOH)_4 + 2\,Ag[(NH_3)_2]OH \longrightarrow (CHOH)_4 + 2\,Ag\downarrow + 3NH_3 + H_2O$$
$$|$$
$$CH_2OH \qquad\qquad CH_2OH$$

with $COO^{\ominus}NH_4^{\oplus}$ on the product chain.

Aldose   **Tollen's Reagent**

Procedure: Take a clean and dry test tube. Add aqueous solution of the given organic compound (0.05 g in 1.0 mL water) followed by addition of freshly prepared Tollens' reagent (2.0 mL). Heat the contents on the water bath for 5 minutes. Appearance of silver mirror on the inner walls of the test tube confirms the presence of reducing sugar.

Note

- The test is highly sensitive. If the test tube is not clean, only black residue will be formed in place of silver mirror.
- Do not heat in boiling water bath as the reagent may decompose to give false positive.
- Non-reducing sugars like sucrose reduce Tollen's reagent upon hydrolysis

Safe Waste Disposal

- See Tollens' test, *section 4.3.2.2.1,* page 151.

(b) *Fehling test or Benedict's test:* Fehling solution is obtained by mixing equal volumes of Fehling solution A and Fehling solution B, whereas Benedict's reagent is a single test solution. Chemically, Benedict's solution is an alkaline solution of citrate. Being, weakly alkaline as compared to Fehling solution, it does not deteriorate on standing. (For details see Fehling solution test, *section 4.3.2.2.1,* page 153).

Reducing sugars produce brick red precipitate of cuprous oxide on heating with Fehling solution or Benedict's solution. The reaction is essentially the same as for aldehydes.

$$\underset{\text{Aldose}}{RCHO} + 2Cu^{2\oplus} + 5OH^{\ominus} \longrightarrow RCOO^{\ominus} + \underset{\text{Brick red ppt.}}{Cu_2O\downarrow} + 3H_2O$$

Procedure: Heat Fehling solution or Benedict's reagent (2.0 mL), taken in a clean test tube, bring to boil and then add an aqueous solution of the given compound (0.1 g in 2.0 mL of water) and heat on a low flame directly on Bunsen burner for another 2-3 minutes. Appearance of brick red or yellow-red precipitate confirms the presence of reducing sugar.

Note

- Normally, non-reducing sugars do not give positive test. However, some non-reducing sugars like sucrose may also give a faint red precipitate due to partial hydrolysis on heating.
- Non-reducing sugars like sucrose, on subjecting to hydrolysis with 10% aqueous hydrochloric acid, followed by neutralization with 10% aqueous NaOH, will give a positive test with Benedict's and Fehling's solutions.
- α-Hydroxy aldehydes, α-hydroxy ketones, and α-keto aldehydes also respond to this test. Aromatic aldehydes are however not oxidized under these conditions.

Safe Waste Disposal

- Filter the red coloured precipitate of cuprous oxide and transfer the residue to the non-hazardous solid waste container. Transfer the filtrate to the aqueous waste solution container.

5. Tests for distinction between aldoses and ketoses: If the given sugar is a monosaccharide then the next step will be to find out whether it is an aldose or ketose. This can be done by doing any one of the following two tests:

(a) *Seliwanoff's test:* It is based on the fact that the ketoses readily dehydrate to form furfural or substituted furfural (HMF) ,when heated with concentrated sulphuric acid, whereas in case of aldoses there is a time lag. Furfural or 5-hydroxymethylfurfural (HMF) formed during the reaction condense with resorcinol, in acidic medium, to form cherry red or blue coloured complex.

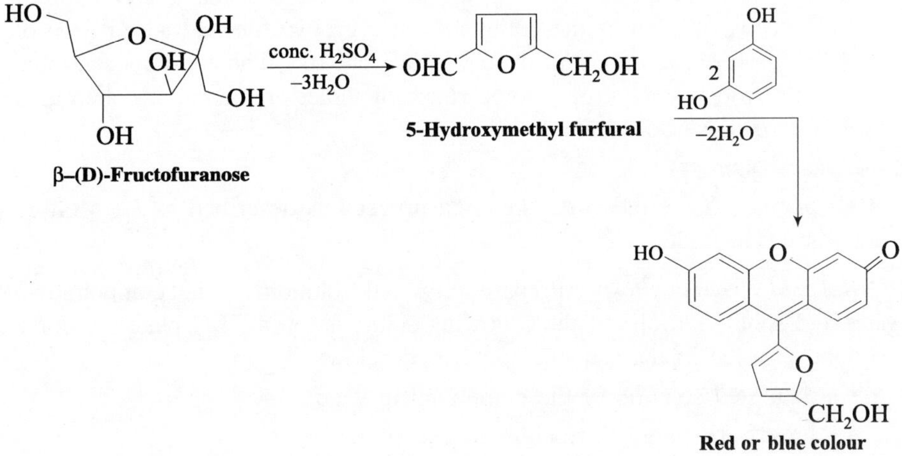

β–(D)-Fructofuranose

5-Hydroxymethyl furfural

Red or blue colour

Procedure: Take aqueous solution of the organic compound (0.1 g in 2.0 mL water) in a clean test tube and add Seliwanoff's reagent (2.0 mL). Appearance of cherry red or blue precipitate indicates the presence of a ketose.

(b) *Rapid furfural test:* The chemistry of rapid furfural test is similar to Molisch test, only concentrated sulphuric acid is replaced by concentrated hydrochloric acid. This test is based on the fact that furanoses rapidly produce violet colouration due to rapid dehydration to 5-hydroxymethylfurfural (HMF) followed by complexation with 1-naphthol (See Molisch test, page 180).

Procedure: Take a dilute solution of glucose (0.05 g in 1.0 mL water) in a clean test tube, add 10% alcoholic solution of 1-naphthol (1.0 mL), add concentrated HCl (6-8 mL) and boil on a low Bunsen burner flame. Fructose or sucrose produce instant violet colouration, as the solution starts to boil. In case of glucose, violet colouration is produced after boiling for 1-2 minutes.

6. Spot tests for carbohydrates

(a) *Molisch test:* Prepare saturated solution of the compound in water and spot the following solutions (1-2 drops) on a TLC plate or Merck strip, in the sequence described below:

- Saturated solution of the compound in water
- Molisch reagent (2-3 times)
- Concentrated sulphuric acid

Violet or red-violet colouration confirms carbohydrate.

Note

- This test can also be done in a groove tile as follows: Take a drop of aqueous solution of the compound in a groove tile and add a drop of Molisch reagent and mix well with a glass rod. Then add concentrated sulphuric acid (1 drop). Appearance of violet or red-violet colouration confirms carbohydrate.

Safe Waste Disposal

- Scrape off the red-violet spot and proceed as described in the Molisch test, page 180.

(b) *Barfoed's reagent test:* Prepare saturated solution of the compound in water and spot 1-2 drops of the following solutions on a TLC plate or Merck strip and proceed in the sequence described below:

- Saturated solution of the compound in water.
- Freshly prepared Barfoed's reagent
- Heat the strip in the oven maintained at 120°C, for 1-2 minutes

Formation of brick red spot confirms reducing sugar.

Note

- Perform the blank test by spotting a drop of Barfoed's reagent on one side of the paper strip and heating as above.
- Saturated solution of copper acetate in 1% acetic acid may be used as Barfoed's reagent for this test.

Safe Waste Disposal

- Scrape the red spot in a test tube and proceed as described in Barfoed's reagent test, Page 182

(c) *Fehling solution test:* Prepare saturated solution of the compound in water and spot 1-2 drops of the following solutions on a Whatman paper strip and proceed in the sequence described below:

- Saturated solution of the compound in water.
- Freshly prepared Fehling solution (equal volumes of Fehling A and B, spotted thrice).

- Heat the strip in the oven maintained at 120°C for 1-2 minutes.

Formation of brick red spot confirms reducing sugar

Note

- Perform the blank test by spotting a drop each of Fehling's A and B solutions on top of each other on one side of the paper strip and heating as above.

Safe Waste Disposal

- Scrape the red spot in a test tube and proceed as described in Fehling's solution test, page 184.

Derivatives of Carbohydrates

1. Osazone: Sugars like glucose, fructose, mannose and α-hydroxyaldehydes and ketones react with three molar equivalents of phenylhydrazine to form bright yellow crystalline osazone, which is sparingly soluble in water. The reaction involves formation of hydrazone on C-1 of the aldose or C-2 of the ketose. This is followed by Amadori rearrangement to form osazone (Bis-phenylhydrazone). The reaction is arrested at the bis-phenylhydrazone (osazone) stage, due to chelate formation.

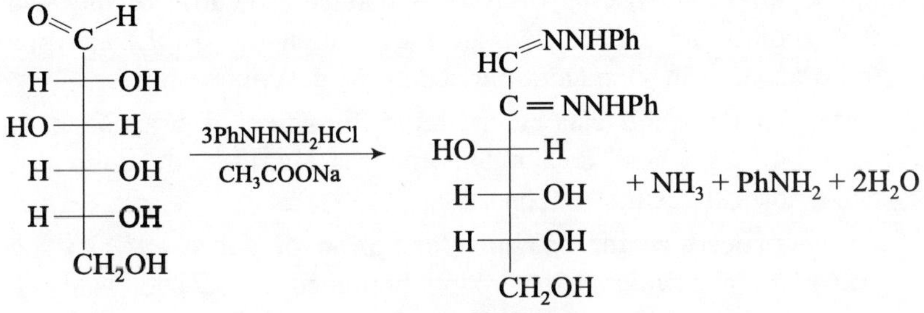

$$\text{Glucose} \xrightarrow[\text{CH}_3\text{COONa}]{3\text{PhNHNH}_2\text{HCl}} \text{Glucosazone} + NH_3 + PhNH_2 + 2H_2O$$

Glucose

Glucosazone

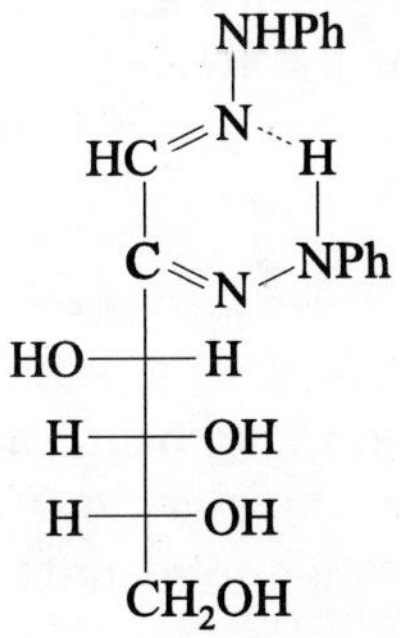

Chelate formation

For detailed theory, see Volume 1, *section 6.18.1.*

Procedure

(a) Dissolve the given reducing sugar (1.0 g) in water (2-3 mL) in a boiling tube. Prepare a solution of phenylhydrazine hydrochloride (2.0 g) and anhydrous sodium acetate (3.0 g) in water (2-3 mL) and add to the sugar solution. Warm until a clear solution is obtained, filter if required. Next, put a glass rod and place a cotton plug on the mouth of boiling tube. Clamp the boiling tube vertically in the water bath such that the base of the boiling tube is dipped inside the water. Heat the contents on boiling water bath for 20-30 minutes. Allow the reaction mixture to attain room temperature, dilute with water (3-4 mL) then cool in ice bath for 10 minutes. Filter the precipitated bright yellow solid. Wash with very dilute hydrochloric acid and then with water. Dry the sample and record its melting point.

Non-reducing sugars like sucrose cannot form osazone but it can be hydrolyzed to equimolar mixture of monosaccharides (glucose and fructose) which are both capable of forming the osazone. Following procedure should be used for the osazone formation of sucrose:

(b) Place sucrose (1.0 g), water (10.0 mL), 1-2 pumice stones and a few drops of concentrated sulphuric acid (AR) in a round bottomed flask fitted with a reflux water condenser. Reflux the contents on water bath for 30 minutes. After hydrolysis is complete, neutralize the solution with sodium acetate, concentrate the solution (~2.0 mL) and prepare the osazone using the same procedure as described above.

Note: This method can be used for formation of osazone of those disaccharides whose constitutent monosaccharides are epimeric and hence form the same osazone.

(c) **Green procedure for osazone formation of sucrose:** Take sucrose (1.0 g) in a clean and dry round bottomed flask, add water (10-15 mL) and then add dilute HCl (5.0 mL). Heat at ~70°C, in a microwave oven for 5 minutes. Cool and add phenyl hydrazine hydrochloride and sodium acetate as detailed in (a). Heat on the microwave oven for 2 minutes and work up as in (a).

Note

- Osazone of some sugars like lactose separate only when heating for 30 minutes is followed by cooling in ice-water bath.

- Time required for osazone formation on boiling water bath varies from sugar to sugar. For example, fructose (2 minutes), glucose (10-15 minutes). Note the time required for osazone formation to identify the probable sugar present.

- Most osazones darken and decompose just before melting. This should not be confused with the melting point of the compound.
- Acetic acid liberated in the reaction helps to catalyse the formation of osazone.
- The green process is faster and consumes less energy.

Safe Waste Disposal

- Place the solid product in a non-hazardous organic waste container. Add sodium hypochlorite (5.25%, 8.0 mL) to the filtrate, heat at 45-50°C for 2 hours to oxidize phenyl hydrazine. Dilute with 10.0 mL of water and transfer to the aqueous waste container.

2. Acetate: Carbohydrates react with excess of acetic anhydride, in the presence of a suitable catalyst to form the corresponding acetyl derivative. For example, glucose forms pentaacetyl derivative which can exist in two anomeric forms corresponding to α- and β- D(+)-glucose. α- and β-Anomers can be isolated using zinc chloride and sodium acetate, as catalyst, respectively. It is likely that the product in each case is contaminated with traces of other anomer, which can be removed easily by recrystallization.

$$\text{HO} \underset{\text{OH}}{\overset{\text{OH}}{\text{Glucose ring}}} \xrightarrow[\text{Anhy. ZnCl}_2]{(CH_3COO)_2O \text{ (5 moles)}} \text{H}_3\text{COCO} \underset{\text{OCOCH}_3}{\overset{\text{OCOCH}_3}{\text{ring}}}$$

α-D(+)-Glucose → α-D(+)-glucose pentaacetate

Procedure

(a) *α-Acetate:* Take zinc chloride (2.0 g) and fuse it in a clean and dry china dish (See Appendix II for details). Transfer fused zinc chloride to a clean and dry round bottomed flask fitted with a reflux water condenser, add acetic anhydride (10.0 mL) and 1-2 pumice stones. Reflux the contents vigorously on water bath for 5 minutes or till zinc chloride dissolves completely. Remove the condenser and add the given carbohydrate (1.0 g). Shake the mixture and put the reflux condenser and reflux the contents on the boiling water bath for one hour. Pour the contents with stirring into ice-cold water (100 mL). Allow the mixture to stand for 10 minutes. White coloured crystals of acetate derivative separate out. Filter, wash with water (5-7 mL), and dry. Recrystallize from ethanol and determine the melting point.

(b) *β-Acetate:* Mix acetic anhydride (8.0 mL), anhydrous sodium acetate (1.5 g) and 1-2 pumice stones in a dry round bottomed flask fitted with a water condenser. Heat the contents on water bath for 5 minutes to ensure complete dissolution of sodium acetate. Add the given carbohydrate (1.0 g) to the above mixture and reflux for 30-120 minutes. Cool and transfer the contents to ice-cold water (50-70 mL), with stirring. Colourless crystals of β-acetate separate on filtration at suction, wash well with water and dry the sample. Recrystallize a fraction of the crude sample from ethanol and determine its melting point.

Note

- Before filtering, ensure that the reaction mixture does not smell of acetic anhydride.

- Carry out acetylation to completion otherwise the melting point of the synthesized product does not tally the literature value.

- Check for completion of reaction by pouring 1-2 drops of the reaction mixture in ice-water. Look for colourless precipitate, if formed, the reaction is ready for workup.

- Always use freshly fused sodium acetate or zinc chloride. (See Appendix II)

- Fuse zinc chloride and sodium acetate on a low flame to avoid charring.

- Heating time for β-acetates varies from 30-120 minutes, so check for completion of reaction before work-up.

Safe Waste Disposal

- Transfer the entire aqueous filtrate to the aqueous waste solution container.

Post Lab Questions

1. What is Molisch reagent?
2. What precautions should you take while performing Molisch test?
3. Why do carbohydrates char on warming with concentrated sulphuric acid?
4. Give chemical test/s to distinguish between:
   - (a) Ribose and 2-deoxy ribose (Any one)
   - (b) Glucose and sucrose (any two)
   - (c) Glucose and fructose (any one)
   - (d) Maltose and sucrose (any one)
5. A student prepared glucaosazone derivative and did not wash the crude sample with dilute HCl, followed by water. Do you think he will get the correct melting point of the product? If not, why?
6. Can you identify carbohydrates by their melting points? Explain.
7. Carbohydrates like glucose and fructose:
   - (a) Reduce Tollens' reagent
   - (b) Do not react with 2,4-DNP but react with phenylhydrazine
   - (c) Give positive CAN test.

(d) Give brisk effervescence with sodium metal. Explain all the observations.

8. Name the commonly occurring carbohydrates found in:

   (a) Milk   (b) Tree trunk

   (c) Grapes   (d) Sugar cane

   (e) Rice   (f) Honey

9. What is Barfoed's reagent? Can you use Fehling's solution to distinguish between monosaccharide and disaccharide? If not, why?

10. How is sugar tested in

    (a) Urine   (b) Blood

    Explain the chemistry involved.

11. Glucose and fructose cannot be differentiated by osazone formation. Explain.

12. Explain the role of anhydrous zinc chloride in the acetylation of sugars with acetic anhydride.

4.3.2.3 Compounds Containing Carbon, Hydrogen and Halogen (Neutral Compounds)

These are compounds containing halogen atom attached directly to the aromatic ring, like C_6H_5Cl, C_6H_6Br, and C_6H_6I or halogen atom attached to aromatic side chain like benzyl chloride, $C_6H_5CH_2Cl$ or compounds of type RX where, R is aliphatic. All these compounds are liquid, insoluble in water and settle at the bottom, being heavier than water, unlike other organic compounds. Aryl halides like chlorobenzene, bromobenzene and iodobenzene possess aromatic odours whereas arylalkyl halide like benzyl chloride ($C_6H_5CH_2Cl$) is lachrymatory and has a sharp irritating odour.

Halogen atom attached to an organic compound does not impart any characteristic solubility to the parent compound. Therefore, the presence of halogen in the organic compound is immaterial as far as the solubility classification is concerned. However, displacement reaction of these neutral halohydrocarbons with alcoholic silver nitrate or sodium iodide in acetone may be used to further categorise the type of halohydrocarbons. The classification tests for identification of halohydrocarbons are listed in Table 4.20.

Table 4.20 Classification Tests for Halohydrocarbons

S. No.	Classification test	Test category	Observation	Inference
1.	Lassaigne's Sodium Test for halogens	Preliminary test	White/pale yellow/ deep yellow precipitate	Halogen confirmed*
2.	Layer test	Special test	Orange/Violet coloration in organic layer	Bromine/Iodine confirmed

Contd.

3.	2% Alcoholic silver nitrate test	Special test	White or yellow precipitate	Tertiary or secondary alkyl halide/aryl-alkyl halide confirmed*
4.	NaI/Acetone	Special test	White or yellow precipitate	Primary alkyl/aryl-alkyl bromide or chloride confirmed

*See test for halogens, *section 4.3.1.6.4*, page no. 111, for details.

Note

- Perform the Beilstein test before carrying out the Lassaigne's sodium test for halogens.

- After obtaining positive Beilstein test and Lassaigne's sodium test, if test for nitrogen and all other functional group tests discussed in *section 4.3.2.1 and 4.3.2.2* are negative, the compound is safely assumed to be halohydrocarbon. Proceed with the following tests for further characterisation of the compound.

Classification Tests for Halohydrocarbons

1. 2% Alcoholic silver nitrate test: This test can be used for distinguishing aryl halides from aryl-alkyl halides or simple tertiary or secondary alkyl halides from primary alkyl halides. The reaction proceeds *via* S_N1 mechanism. Therefore, all those halohydrocarbons capable of forming stable carbocation will respond to this test.

Order of reactivity: $R_3C\text{-}X > R_2CH\text{-}X > RCH_2\text{-}X$; **Mechanism:** S_N1

$$R_3CX \;+\; AgNO_3 \xrightarrow{\text{Alcohol}} R_3CONO_2 \;+\; AgX\downarrow$$

3° Halide   Silver nitrate   Silver halide
                                                                             White or yellow precipitate

Procedure: See *section 4.3.1.6.4*, page 111, for details.

Safe Waste Disposal

- Saturate the aqueous solution with solid sodium chloride. Remove the precipitate of silver halide by filtration and place in the non-hazardous solid waste container. Place the aqueous filtrate in the aqueous solution container.

2. Sodium iodide in acetone test (Finkelstein Test):   This qualitative test is used to categorise alkyl chlorides and bromides as primary, secondary, or tertiary. It is based on the fact that sodium chloride and bromide is partially soluble in acetone whereas sodium iodide is completely soluble. The reaction involves replacement of chlorine or bromine in alkyl halides by iodine to produce a white/yellow coloured precipitate of sodium chloride/bromide. The reaction proceeds *via* S_N2 displacement and therefore the expected order of reactivity of alkyl halides will be:

Primary > Secondary> Tertiary; Br>Cl (Leaving group ability)

$$RCl \quad + \quad NaI \quad \xrightarrow{\text{Acetone}} \quad RI \quad + \quad NaCl\downarrow$$

Alkyl chloride **Sodium iodide**   **Alkyl iodide** **Sodium chloride**

$$RBr \quad + \quad NaI \quad \xrightarrow{\text{Acetone}} \quad RI \quad + \quad NaBr\downarrow$$

Alkyl bromide **Sodium iodide**   **Alkyl iodide** **Sodium bromide**

Procedure: Take a 15% sodium iodide reagent (1.0 mL, see Appendix II) in a clean test tube and add the given organic compound (2-drops or 0.1 g of the compound in minimum amount of acetone). Mix the contents and allow the test solution to stand at room temperature (~3 min). Note if any precipitate is formed. If no change occurs at room temperature, place the test tube in a beaker of water at 50°C, warm for ~6 min and then cool. Appearance of yellow or white precipitate confirms the presence of alkyl bromides or chlorides respectively.

Note

- Ease of substitution of halogens follows the order: **Iodine>bromine>Chlorine**
- Displacement reactions in alcoholic silver nitrate and NaI/acetone are complementary and are thus used to categorize alkyl /aryl-alkyl halides as primary/secondary/tertiary.
- Primary alkyl bromides will react faster with Finkelstein's reagent than corresponding chlorides, which require heating at 50°C for 5-6 minutes. Secondary and tertiary bromides also require heating at 50-60°C for ~6 minutes, whereas tertiary chlorides do not react at all. These results are in line with the S_N2 mechanism and form the basis of distinction between the different classes of alkyl bromides and chlorides.
- Benzyl and allyl halides react equally well with alcoholic silver nitrate to give a positive test due to formation of stable carbocation.
- Aryl and vinyl halides do not show any of these displacement reactions because resonance increases the stability of the C-X bond. Some substituted aryl halides, having electron withdrawing groups, may respond to these test.
- In Finkelstein test, excessive heating in water bath may lead to evaporation of acetone and precipitated NaI may give a false positive test.
- Carboxylic acids precipitate silver salts on reaction with silver nitrate. However, these silver salts are soluble in dilute nitric acid, while silver halides are not.

Safe Waste Disposal

- Transfer the test solution to the aqueous waste solution container.

3. Spectroscopic Identification

Table 4.21   Spectral Data of Halohydrocarbons

S.No.	Type of Spectrum	Peak value	Peak attribution	Remarks
1.	IR	1035-1100 cm^{-1} 1030-1075 cm^{-1}	C-Cl Str. (Aromatic) C-Br Str. (Aromatic)	- -
2.	NMR	2-4 δ*	R$\underline{H}_2$X (X=Cl, Br, I)	-

*δ values increase with the increase in the number of halogen atoms. For example, chloroform appears at 7.2δ whereas dichloromethane appears at 5.3δ.

Derivatives of Aromatic Halohydrocarbons

1. Nitro derivative: Aryl halides undergo nitration with 1:1 mixture of concentrated/fuming nitric acid and concentrated sulphuric acid (nitrating mixture). The nature of the product formed depends on the amount of nitrating mixture used and also on the nature of the substrate. For example, reaction of chlorobenzene with one mole equivalent of nitrating mixture will yield 4-chloronitrobenzene whereas reaction with two mole equivalents produces 1-chloro-2,4-dinitrobenzene.

Cl

conc. HNO₃

conc. H₂SO₄

Cl

NO₂

NO₂

Chlorobenzene   **1-Chloro-2,4-dinitrobenzene**

Procedure: Prepare nitrating mixture by placing concentrated nitric acid (4.0 mL) in a clean and dry round bottomed flask, cool in ice bath. Add concentrated sulphuric acid (4.0 mL) dropwise with shaking. To this solution, add the unknown halohydrocarbon (1.0 mL), in small lots, with constant shaking. Once the addition is over and the exothermic reaction has subsided, reflux the contents on water bath for 15-30 minutes. Allow the contents to cool to room temperature and then pour into ice-cold water (50.0 mL). Filter the separated solid, wash with water and dry. Recrystallize from dilute ethanol and determine the melting point.

Note

- This procedure will lead to dinitro derivative for aryl halides like chlorobenzene.

- Concentrated acids are highly toxic and corrosive. Read MSDS before handling and disposal (https://Pubchem.ncbi.nlm.nih.gov).

- Check for completion of reaction before workup. This can be done by adding 1-2 drops of the reaction mixture to ice-cold water taken in a boiling tube. Appearance of solid indicates that the reaction is ready for workup.

- Note the bitter almond smell in the products.
- Nitration of unknown aromatic compound should be carried out cautiously because many of them behave violently under the reaction conditions. Try a small test tube sample before carrying out the reaction on large scale.
- In certain cases, concentrated nitric acid may be replaced by fuming nitric acid (4.0 mL) and concentrated sulphuric acid by 20% oleum, if nitration does not take place under the conditions described above.
- Temperature required for nitration depends on the nature of the substrate and type of the nitric acid used (fuming or concentrated).
- In some cases, initially an oil separates on workup, but it solidifies on cooling and scratching the sides of beaker with glass rod.

Safe Waste Disposal

- Transfer the aqueous filtrate to a 250 mL beaker containing water (20.0 mL). Add solid sodium carbonate until foaming ceases and pH ~7. Transfer the solution to the aqueous waste container.

Post Lab Questions

1. Monochloro ethers like, $ClCH_2OR$, are highly reactive to $AgNO_3$. Explain.
2. Benzyl chloride reacts much faster with sodium iodide in acetone than cyclohexylmethyl chloride, even though both are primary alkyl halides, explain.
3. Explain why benzyl chloride is reactive towards both ethanolic silver nitrate as well as NaI in acetone, whereas chlorobenzene is unreactive.
4. Tertiary butylchloride does not give Finkelstein test. Explain.

4.3.2.4 Compounds Containing Carbon and Hydrogen only (Neutral Compounds)

If your preliminary tests suggest that your compound is aromatic or unsaturated and does not respond to any other functional group test or elemental analysis, you are safe to conclude that your compound is a hydrocarbon. Unsaturated hydrocarbons and some arenes are sulphonated at room temperature, thus can be easily identified by their solubility in concentrated sulphuric acid (No excessive charring is observed!). Further, IR and NMR spectra give clear signals for the presence of unsaturation and aromatic nucleus in the organic molecule.

Benzene and its derivatives are mostly colourless liquids, insoluble in water and float on top, being lighter than water. Naphthalene, $C_{10}H_8$, is a colourless solid, melting point 80°C and has characteristic moth ball odour. Anthracene is also a colourless solid, melting point 216°C, if pure, but slowly turns pale yellow. Biphenyl, m.p. 69°C, and phenanthrene, m.p. 98°C are colorless solids.

Alkanes being unreactive are best identified by spectroscopic peaks. Unsaturated hydrocarbons are identified by bromine solution and Baeyer's reagent tests (See *section 4.3.1.8*, page 116, for details). A list of classification tests for aromatic hydrocarbons is given in Table 4.22.

Table 4.22   Classification Tests for Aromatic Hydrocarbons

S. No.	Classification test	Test category	Observation	Inference
1.	Solubility in concentrated sulphuric acid	Preliminary test	Soluble on warming and no appreciable charring is observed.	Possibly an aromatic hydrocarbon*
2.	Solubility in 20% fuming sulphuric acid**	Special test	Clear solution on warming	Benzene, toluene, naphthalene, etc.
3.	Chloroform and AlCl₃ test	Special test	Red, blue, green or purple colouration	Aromatic hydrocarbon

* Alcohols, carbonyl compounds and unsaturated hydrocarbons also respond to this test. Benzene, toluene and naphthalene do not give this test. Check the flame test, solubility in other solvents and functional group tests before arriving at any conclusion.

**To be done for compounds giving negative functional group test, lacking nitrogen and not responding to solubility in concentrated sulphuric acid. For example, benzene, toluene and naphthalene.

Chemical tests for aromatic hydrocarbons

1. Solubility in concentrated sulphuric acid: See *section 4.3.1.5*, page 98, for solubility in concentrated sulphuric acid.

2. Solubility in 20% fuming sulphuric acid: Aromatic hydrocarbons dissolve in fuming sulphuric acid due to formation of soluble arene sulphonic acids.

$$\underset{\textbf{Aromatic compound}}{ArH} \xrightarrow[SO_3]{H_3SO_4} \underset{\textbf{Arylsulphonic acid}}{ArSO_3H} + Heat$$

Procedure: Place 20% fuming sulphuric acid (0.5 mL) in a clean, dry test tube and add the unknown compound (0.25 mL or 0.25 g). Mix and allow the contents to stand for a few minutes. Complete dissolution of the compound with evolution of heat but no appreciable charring, confirms aromatic hydrocarbon.

Note

- Carry out the reaction in fume hood.
- Fuming sulphuric acid being a stronger sulphonating agent, sulphonates naphthalenes, benzenes at room temperature while concentrated sulphuric acid does not.
- Saturated aliphatic hydrocarbons do not dissolve even in fuming sulphuric acid. Aromatic sulphonic acids are water soluble, hence do not precipitate on dilution.
- This test should be done only if the compound does not contain nitrogen or sulphur.

- Compounds containing OH or NH_2 group, attached to the aromatic ring, react violently with fuming sulphuric acid. This test should not be performed on these compounds.

Safe Waste Diposal

- In the fume hood, carefully transfer the test solution into a beaker containing 10.0 mL of water. Add solid sodium carbonate, in small lots (caution: foaming!) until foaming stops. Transfer the test solution to the aqueous waste solution container.

3. Chloroform and aluminium chloride test: Aromatic compounds condense with chloroform and aluminium chloride (Friedel Crafts reaction) to produce triaryl carbocation, which remains in solution as coloured tetrachloroaluminate salts. No reaction is observed for aliphatic hydrocarbons.

$$3ArH \ + \ CHCl_3 \ \xrightarrow{AlCl_3} \ Ar_3CH \ + \ 3HCl$$

Aromatic compound   **Triarylmethane**

$$Ar_3CH \ + \ AlCl_3 \ + \ 2Cl^{\ominus} \ \longrightarrow \ [Ar_3C]^{\oplus} \ AlCl_4^{\ominus} \ + \ HCl$$

Tetrachloroaluminate
(coloured)

Procedure: Take the unknown compound (0.1 mL or 0.1 g) in a clean and dry test tube and add dry chloroform (2.0 mL). Mix thoroughly, and incline the test tube so as to moisten the wall. Then add anhydrous aluminum chloride (0.5-1.0 g) so that some of the powder strikes the side of the moist test tube. Note the colour of the powder on the side, as well as in solution. Bright coloured solution indicates aromatic compound (See Table 4.23).

Safe Waste Disposal

- Place the test solution in the halogenated organic waste container.

Table 4.23 Colouration of some Aromatic Compounds with Chloroform and $AlCl_3$

S. No.	Compound	Colour
1.	Naphthalene	Blue
2.	Biphenyl	Purple
3.	Phenanthrene	Purple
4.	Anthracene	Green
5.	Aryl halides or benzene or its derivatives	Orange-Red

Note

- Many aromatic compounds respond to this test. Some may require heating, specially those having electron withdrawing groups.

- Aromatic esters, ketones, amines and other oxygen/nitrogen containing compounds give green colour. However, all these compounds can be identified by their respective functional group test. Hence, should not create confusion.

- This test should be performed only for compounds insoluble in concentrated sulphuric acid.

4. Spectroscopic Identification

Table 4.24   Spectral Data of Hydrocarbons

S. No.	Type of spectrum	Peak value	Peak attribution
1.	IR	3000-3100 cm^{-1}	C-H str. (Aromatic)
		<3000 cm^{-1}	C-H str. (Aliphatic)
		3300 cm^{-1}	C-H str. (terminal alkyne)
		2100-2260 cm^{-1}	C $\equiv$ C (Str., terminal alkyne)
		1500-1600 cm^{-1}	Aromatic (C=C)
		1650 cm^{-1}	C=C str. (unsymmetrical alkenes only)
2.	NMR	6-9 δ	Aromatic – H
		2-3 δ	Ar-CH$_2$–
		4.5-7 δ	Vinylic – H
		1.7-2.5 δ	Allylic – H
		2-3 δ	–C $\equiv$ C – H
		0.9 δ	–CH$_3$(alkyl)

Derivatives of aromatic hydrocarbons

1. Picrate: Many polynuclear hydrocarbons and tertiary amines form 1:1 adduct (π-complexes) with picric acid, in organic solvents like ethanol or acetone. For example, naphthalene reacts with one mole equivalent of picric acid to form naphthalene picrate.

Naphthalene + Picric acid → Naphthalene picrate

The picrates are highly stable, crystalline solids characterized by sharp melting point. Some picrates like anthracene picrate are unstable, therefore they should not be crystallized especially with alcoholic solvents. They can however be safely washed with little ether and dried.

Procedure

(a) *Naphthalene picrate:* Prepare concentrated aqueous solutions of picric acid and naphthalene in cold acetone (2.0 mL, each). Mix the two solutions in a clean and dry china dish. Stir and allow the contents to stand for 10-15 minutes. Long, yellow needle-like crystals of naphthalene picrate will separate out as the solvent evaporates. Recrystallize the residue from ethanol.

(b) *Anthracene picrate:* In a clean and dry china dish, prepare concentrated solution of anthracene in hot acetone (2.0 mL). Add to this solution, dropwise a concentrated solution of picric acid in cold acetone (2.0 mL). Warm the contents over steam water bath for 1-2 minutes, stir and allow the contents to stand at room temperature for 10-15 minutes. Ruby red crystals of anthracene picrate will separate out as acetone evaporates. Filter, dry and record melting point of crystals.

Note

- Avoid adding excess of picric acid solution as the red colored crystals of anthracene picrate may become contaminated with yellow picric acid.

- Avoid using ethanol as solvent for anthracene picrate as the latter is unstable in hydroxylic solvents. Ethanol can however be used as solvent for picrates of naphthalene and other hydrocarbons.

- Picric acid stains skin yellow, avoid direct contact. Always use gloves.

- Picrates of benzene and toluene are difficult to prepare, should therefore be avoided. Nitration and side chain oxidation, as applicable, may be a better method in these cases.

Safe Waste Disposal

- Place all the test solutions in non-hazardous organic waste container.

2. Nitration of aromatic hydrocarbons: Nitration of aromatic hydrocarbons can be carried out with either 1:1 mixture of concentrated nitric acid and concentrated sulphuric acid or by replacing concentrated nitric acid by fuming nitric acid. (For detailed theory, see Volume 1, *section 6.4*).

$$\text{Benzene} \xrightarrow[\text{conc. } H_2SO_4]{\text{conc. } HNO_3} \text{m-Dinitrobenzene}$$

Benzene   **m-Dinitrobenzene**

Procedure: Prepare nitrating mixture by placing concentrated nitric acid (4.0 mL) in a clean and dry round bottomed flask, fitted with a reflux condenser. Cool in ice. Add concentrated sulphuric acid (4.0 mL) dropwise with shaking.

To this solution, add the unknown hydrocarbon (1.0 mL) in small lots with constant shaking. Once the addition is over and the exothermic reaction has subsided, reflux the contents on wire gauze for 30-45 minutes. Check for completion of reaction by transferring a drop of the reaction mixture to ice-water taken in a beaker. Appearance of yellow precipitate indicates completion of reaction. If the reaction is complete, stop heating and allow the contents to cool to room temperature, else allow the reaction to proceed to completion, and then cool. Transfer the cooled contents, with stirring, into ice-cold water (40-50 mL) in a 100 mL beaker. Allow the reaction mixture to stand in ice-water bath for 5-10 minutes. Filter the separated solid, wash with water, and dry. Recrystallize from dilute ethanol and determine its melting point.

Note

- This procedure will lead to dinitro derivative. For mononitration of the given hydrocarbon, the amounts of concentrated nitric acid and concentrated sulphuric acid need to be reduced to half.

- Fuming nitric acid (3.0 mL) can replace concentrated nitric acid in the above procedure. Reaction in such cases proceeds smoothly in boiling water bath.

- Mono nitration of more reactive polynuclear hydrocarbons can be done as follows: Dissolve naphthalene (1.0 g) in glacial acetic acid (5.0 mL), by gentle warming. Add fuming nitric acid (1.0 mL) and heat at 80°C for 1 minute. Pour the clear yellow solution into ice-cold water (50 mL) with stirring. Yellow crystalline 1-nitronaphthalene will separate on standing. Filter the solid, wash well with water and dry. Recrystallize a part of the precipitate from ethanol.

- Nitrophenols and nitroamines cannot be nitrated by this method. The highly activated ring will undergo oxidation in such cases. OH or NH_2 group in such compounds can be exploited for derivatization.

- Nitro derivatives of certain aromatic compounds, for example, toluene, nitrotoluenes, *m*-dinitrotoluene are highly explosive. Polynitration should be avoided in such cases.

- Also see nitration of halohydrocarbons (*section 4.3.2.3*, page 194).

Safe Waste Disposal

- See nitration of halohydrocarbons, *section 4.3.2.3*, page 194.

3. Side chain oxidation: Aromatic hydrocarbons with side chain like toluene and xylenes undergo oxidation to form aromatic acids. Benzoic acid can be obtained by oxidation of toluene, benzyl alcohol or any other monoalkyl benzene having at least one benzylic hydrogen. The oxidation proceeds easily in alkaline or acidic potassium permanganate or acidic potassium dichromate. The higher reactivity of benzylic hydrogen as compared to other hydrogens in the alkyl side chain forces the oxidation reaction to proceed at this site only. Hence, all alkylbenzenes irrespective of their chain length give benzoic acid as a product under these reaction conditions. For example, toluene can be oxidized to benzoic acid using alkaline potassium permangante. (For detailed theory see oxidation, *section 6.11.1*, Volume 1).

$$CH_3\text{-benzene} + 3[O] \xrightarrow[\text{2. Dil.HCl}]{\text{1. Alk.KMnO}_4} COOH\text{-benzene} + H_2O$$

Toleune   **Benzoic acid**

Procedure: Take a mixture of unknown hydrocarbon (1.0 mL), sodium carbonate (1.0 g) and water (15.0 mL) in a round bottomed flask fitted with a reflux condenser. Add a saturated solution of potassium permanganate (1.5 g) in minimum amount of water (~5-7 mL). Add 1-2 pumice stones and reflux the contents on wire gauze for 3-4 hours or till the oily droplets persist. Allow the contents to cool down to room temperature, filter the precipitated manganese dioxide and transfer the filtrate to a 250 mL beaker. Acidify with concentrated hydrochloric acid (caution: fumes) and then add solid sodium sulphite in small lots, with stirring, till the brown precipitate of any residual mangenese dioxide dissolves completely. Allow the solution to cool to room temperature and filter the separated colourless crystals of carboxylic acid. Wash well with minimum amounts of cold water and dry the sample. Recrystallize a small part of the crude sample with hot water or aqueous ethanol. Report the melting point of the recrystallized product.

Note

- Time required for refluxing varies from one substrate to another.
- This method is more suitable for oxidation of mono/dialkyl benzenes.
- Aromatic compounds with electron withdrawing groups are difficult to oxidise. For example, nitrotoluenes. Whereas those with electron donating groups (OH, NH_2) show oxidation of the ring rather than side chain.
- Polynuclear hydrocarbons like naphthalene and phenanthrene undergo ring oxidation under the reaction conditions.
- Do not inhale fumes of sulphur dioxide released during addition of sodium sulphite. They can be extremely harmful.

Safe Waste Disposal

- Neutralize the aqueous filtrate with solid sodium carbonate, until pH~7 and then transfer to aqueous solution container.

4.3.2.5 Compounds Containing Carbon, Hydrogen and Nitrogen (Basic Compounds)

If the given organic compound contains nitrogen and is soluble in dilute hydrochloric acid, it means it is basic. The most important group of basic organic compounds are amines and hydrazine derivatives ($RNHNH_2$ or $ArNHNH_2$). The latter are usually not taken up for analysis at the undergraduate level, hence are not discussed in this section.

4.3.2.5.1 Amines

Amines are derivatives of ammonia where one or more hydrogens have been replaced by an alkyl or aryl group. Depending on the number of hydrogen atoms replaced, amines can be classified as primary (RNH_2), secondary (R_2NH) or tertiary (R_3N). Amines behave as Lewis bases due to the presence of lone pair of electrons on the nitrogen atom. The basic character of amines varies with the number of alkyl or aryl groups attached to nitrogen, steric effect of the alkyl or aryl group and solvation factor which comes into play in solution. Table 4.25, summarizes the basicity of some commonly found amines in the undergraduate lab. The readers must note that the basic character of amines increases with the increasing pK_a values.

Table 4.25 pK_a Values of Some Amines Commonly found in Undergraduate Labs

S. No.	Name	pK_a* values (at 25°C)	Safety symbols
1.	Aniline	4.60	
2.	*p*-Toluidine	5.10	
3.	*N*-Methylaniline	4.85	
4.	*N,N*-Dimethylaniline	5.07	
5.	*p*-Nitroaniline	1.01	
6.	1-Naphthylamine	3.92	
7.	2-Naphthylamine	4.16	

8.	Benzylamine	9.33	
9.	*p*-Anisidine	5.36	

The lower aliphatic amines are gases or low-boiling liquids, which are soluble in water up to C-6 after which the solubility decreases as the number of carbon atoms increase. Majority of aromatic amines are liquids (**Note:** p-toluidine, 1- and 2-naphthylamine are solids). All amines have fish-like odour, are colourless when pure, but darken rapidly on exposure to air and light, due to aerial oxidation. They must be stored in amber coloured bottles in the lab. Aromatic amines are practically insoluble in water. Most amines are soluble in dilute HCl except the weakly basic compouds like naphthylamines, which are moderately soluble. A list of classification tests for amines is given in Table 4.26.

Table 4.26 Classification Tests for Amines

S. No.	Classification test	Test category	Observation	Inference
1.	Solubility in dilute HCl	Preliminary test	Soluble but precipitates on neutralizing with aqueous NaOH[a]	Amine
2.	Hinsberg test	Special test	White precipitate soluble/insoluble in NaOH or clear solution on adding dilute HCl	Primary/secondary/tertiary amine
3.	Nitrous acid test	Group test	Clear solution with or without evolution of N_2 gas in cold/yellow oil/colorless or orange colored solution	Primary/secondary/tertiary aliphatic or aromatic amine
4.	Carbylamine test	Special test	Offensive odour of carbylamines[b]	Primary amine (aliphatic/aromatic)
5.	5-Nitrosalicylaldehyde test	Special test	Precipitate not turbidity	Primary amine (aliphatic/aromatic)
6.	Rimini's test	Special test	Red-violet coloured complex	Primary aliphatic amine
7.	Azo dye test	Special test	Orange-red dye	Primary aromatic amine
8.	Liebermann test	Special test	Yellow oil, obtained in nitrous acid test, gives red, blue or green colouration	Secondary amine

Contd.

9.	Nitrous acid test	Special test	Green solid on neutralization of orange solution with NaOH	Tertiary aromatic amine

a: Provided the amine is insoluble in water; **b:** This test should ideally be avoided in the lab.

Note

- If the preliminary tests indicate that the compound contains nitrogen and is soluble in dilute hydrochloric acid then proceed with the classification tests for amines.
- Primary and secondary amines like alcohols, phenols and carboxylic acids give effervescence with sodium metal. This test is however not included in the list as other tests with more concluding evidence are available.
- For elemental analysis, see *section 4.3.1.6*, page 101, and for solubility in dilute HCl, *section 4.3.1.5*, page 98.

Classification tests for amines

1. Hinsberg's test: It is a visual test used for distinction among primary, secondary and tertiary amines. This test does not distinguish an aliphatic and aromatic amine. It is based on the fact that primary amines react with Hinsberg's reagent (benzenesulphonyl chloride) to form a sulphonamide which is soluble in NaOH, due to presence of an acidic hydrogen, but precipitates on acidification. Secondary amines form sulphonamides which are insoluble in NaOH due to lack of acidic hydrogen, and show no change on acidification. Tertiary amines do not form sulphonamides but give clear solution on acidification due to formation of soluble sulphonic acid and amine salt.

$$RNH_2 + PhSO_2Cl + 2NaOH \longrightarrow PhSO_2\overset{\ominus}{NR}\,\overset{\oplus}{Na} + NaCl + 2H_2O$$

1° Amine   Benzenesulphonyl chloride   Soluble

$$\downarrow H^{\oplus}$$

$$PhSO_2NHR$$

Sulphonamide (Insoluble)

$$R_2NH + PhSO_2Cl + NaOH \longrightarrow PhSO_2NR_2 + NaCl + H_2O$$

2° Amine   Benzenesulphonyl chloride   Insoluble

$$\downarrow H^{\oplus}$$

No Reaction

$$R_3N + PhSO_2Cl + 2NaOH \longrightarrow PhSO_3Na + NaCl + R_3N + H_2O$$

3° Amine   Benzenesulphonyl chloride   Insoluble

$$\downarrow HCl$$

$$R_3\overset{\oplus}{N}H\,\overset{\ominus}{Cl} + PhSO_3H$$

Quaternary Ammonium salt (soluble)

Procedure: Take the unknown organic compound (0.3 mL or 0.3 g) in a clean conical flask fitted with a cork, add 10% aqueous sodium hydroxide solution (5.0 mL) and benzenesulfonyl chloride (0.4 mL). Stopper the flask and shake the contents vigorously for 5-6 minutes. Cool the solution after disappearance of Hinsberg's reagent. If clear solution is obtained, acidify by adding 10% aqueous hydrochloric acid. Appearance of white precipitate confirms presence of primary amine. If, however, a precipitate is seen in the original solution after shaking and no change is observed on acidification, the amine is secondary. If the residue remains after shaking with Hinsberg's reagent, treat the solution with 10% aqueous hydrochloric acid. If the residue dissolves on acidification, the amine is tertiary.

Note

- Use pH paper to ensure that the test solution is alkaline during formation of sulphonamide. If not, add some more NaOH solution.
- If the original compound is amphoteric, like amino acid, then Hinsberg's test fails to distinguish between the primary and secondary amine as secondary amine sulpnonamides will also dissolve in NaOH solution due to formation of carboxylates and acidification will precipitate the free acid. For example:

COOH · · · SO$_2$Cl · · · COONa · · · $+$ NaCl $+$ 2H$_2$O

4-(N-Methylamino) benzoic acid · · · Benzenesulphonyl chloride · · · Benzenesulphonamide derivative

Safe Waste Disposal

- Filter the test solution and transfer any residue to the organic non-hazardous solid waste container. If the unknown compound is found to be a primary or secondary amine, then dilute the filtrate with water and transfer the contents to the aqueous waste solution container. If the unknown compound is found to be a tertiary amine, then basify the filtrate with 10% aqueous sodium hydroxide, extract the amine with petroleum ether. Transfer the organic layer to the organic solvents container and put the aqueous layer in the aqueous waste solution container.

2. Nitrous acid test: The reaction of amines with nitrous can be used to not only classify amines as primary, secondary and tertiary but also as aliphatic or aromatic amine. This test can be followed by individual confirmatory tests to confirm primary, secondary or tertiary amines. (see Table 4.27 for details)

$$NaNO_2 + HCl \longrightarrow HONO + NaCl$$
Sodium nitrite

$$RNH_2 \xrightarrow{NaNO_2 + HCl} R - \overset{\oplus}{N} \equiv NCl^{\ominus} \xrightarrow{H_2O} RCl + ROH + ROR + Alkenes + N_2$$
1° Alphatic amine   Diazonium salt

$$ArNH_2 \xrightarrow[0\text{-}5°C]{NaNO_2 + HCl} Ar - \overset{\oplus}{N} \equiv NCl^{\ominus} \xrightarrow[\text{Heat}]{H_2O} HCl + ArOH + N_2$$
1° Aromatic amine   Diazonium salt

$$R_2NH + HONO \longrightarrow \begin{matrix} R \\ \diagdown \\ \diagup \\ R \end{matrix} N - N = O + H_2O$$
2° Amine   Nitrous acid

R = Aliphatic or aromatic   N-Nitrosoamine
(Yellow oil/solid)

$$R_3N + HCl \longrightarrow R_3\overset{\oplus}{N}HCl^{\ominus}$$
3° Aliphatic amine   Soluble

3° Aromatic amine $\xrightarrow[\text{(0-5°C)}]{NaNO_2 / HCl}$ Hydrochloride salt of C-Nitrosoamine (Orange colouration/ppt.)

Procedure: Take a clean and dry test tube. Add the given organic compound (0.5 mL or 0.5 g) followed by dilute hydrochloric acid (1.5 mL concentrated HCl diluted with 2.5 mL of water) to the test tube. Shake the contents. Amine dissolves in dilute acid due to salt formation. Allow the solution to cool to 0°C in a beaker of ice-water. In another test tube kept in ice, cool aqueous solution of sodium nitrite (0.05 g in 2.5 mL of water). Add this cold solution dropwise, with shaking, to the cold solution of amine hydrochloride. Continue addition until the reaction mixture gives positive test for nitrous acid. Allow the reaction mixture to stand for 3-4 minutes in ice-water bath. Look for the characteristic changes listed in Table 4.27 and proceed as directed on pages 209-211.

Note

- Always dilute concentrated acid by adding acid dropwise to water, with stirring. Do not attempt the reverse!
- After addition of sodium nitrite solution, place a drop of the test solution on starch-iodide paper; a blue colour indicates the presence of nitrous acid.

Table 4.27 Nitrous Acid Test for Amines

S. No.	Observation	Inference
1.	Clear solution with effervescence due to evolution of nitrogen gas.	Primary aliphatic or aryl-alkyl amine. Confirm by Rimini's test.
2.	Clear solution with no evolution of gas at low temperature. Warming however, liberates nitrogen gas.	Could be primary aromatic amine. Confirm by Azo dye test.
3.	Yellow-orange oil with no evolution of gas.	Could be secondary aliphatic or aromatic amine. Confirm by Liebermann nitrosoamine test.
4.	Clear solution with immediate positive test for nitrous acid and no evolution of gas. Basification regenerates amine as oily droplets/solid.	Tertiary aliphatic amine which forms a soluble salt with nitrous acid.
5.	Deep Orange-red solution or precipitate.	Tertiary aromatic amine, which forms C-nitroso amine hydrochloride. Confirm by basification with aqueous NaOH.

Note

- Maintain low temperature, 0-5°C, to prevent decomposition of aryl diazonium salts.
- In case of primary aromatic amines, formation of yellow precipitate indicates incomplete diazotization of amine. The diazonium salt couples with the unreacted amine to form yellow coloured azo dye.
- It is important therefore that the reaction medium is feebly acidic when sodium nitrite solution is added. This enables formation of diazonium salt and prevents coupling of the diazonium salt with the amine, by forming amine salt, which remains in equilibrium with the free base.
- Active methylene compounds combine with nitrous acid to form oximes.

3. Confirmatory tests for Primary Amines

(a) *Carbylamine test:* This test is a general test given by all aliphatic or aromatic primary amines. It is based on the fact that primary amines react with chloroform in the presence of alcoholic sodium hydroxide or potassium hydroxide to produce offensive/nauseating odour due to liberation of carbylamine or alkylisocyanide gas.

$$\underset{1^\circ \text{ amine}}{RNH_2} + CHCl_3 + 3KOH \longrightarrow \underset{\text{Isocyanide}}{RNC} + 3KCl + 3H_2O$$

Procedure: Take chloroform (2-3 drops) in a clean and dry test tube. Add the given compound (0.05-0.1 g or 2-3 drops, if liquid) and ethanolic KOH (1.0 mL, 0.5 M) solution. Mix well and heat to boiling in a water bath. The foul/nauseating odour of isocyanide (carbylamine) confirms primary amine. Cool the test tube immediately and add carefully an excess of concentrated hydrochloric acid until the foul odour is no longer noticeable.

$$RNC + H_2O \xrightarrow{H^{\oplus}} [RNHCHO] \xrightarrow{H_2O} RNH_2 + HCOOH$$

Note

- This test is normally avoided in the lab due to the offensive odour of carbylamine. If essential, use fume hood.
- Aqueous HCl hydrolysis carbylamines to less offensive amine.
- This test is given by traces of primary amine present as impurity with secondary or tertiary amine. Please analyse all the tests before concluding.
- Isocyanides are highly toxic, consult MSDS, if performing this test. (https://Pubchem.ncbi.nlm.nh.gov)

Safe Waste Disposal

- Transfer all the test solutions to the aqueous waste container.

(b) *5-Nitrosalicylaldehyde test:* This test is used for detecting the presence of primary amines in the lab. It is based on the fact that primary amines react with 5-nitrosalicylaldehyde to form imines which precipitate coloured complex with nickel chloride.

RNH$_2$ + 2 (CHO, OH, O$_2$N — 2-Hydroxy-5-nitrobenzaldehyde) $\xrightarrow{-H_2O}$ 2 (H–C=N–R, OH, O$_2$N) $\xrightarrow{NiCl_2}$ Coloured Solid

1° Amine   2-Hydroxy-5-nitrobenzaldehyde   Coloured Solid

Procedure: Take the given organic compound (one or two drops or 0.05 g) in a clean test tube and add water (5.0 mL) along with 1-2 drops of concentrated HCl to dissolve the compound. Take the amine solution (0.5-1.0 mL) and transfer to a test tube containing nickel chloride and 2-hydroxy-5-nitrobenzaldehyde reagent (3.0 mL, see Appendix II). Primary aliphatic amines immediately form a precipitate whereas primary aromatic amines produce a definite precipitate in 2-3 minutes.

Note

- Slight traces of primary amines might produce a turbidity, which should not be taken as a positive test.
- This test is given by both aliphatic and aromatic primary amines.
- This test is safer than carbylamine test which produces offensive odor of carbylamine.

Safe Waste Disposal

- Filter the solid at suction and transfer it to the hazardous solid waste container. Transfer the filtrate to the hazardous liquid waste container.

(c) *Rimini's test:* This test is a special test for primary aliphatic amines. The test involves formation of an imine between the primary aliphatic amine and acetone. The imine then forms a colored complex with freshly prepared sodium nitroprusside.

Procedure: Take the given compound (one drop) in a clean and dry test tube and add water (1.0 mL) followed by addition of acetone (2.0 mL). Then add freshly prepared 1% sodium nitroprusside solution (1.0 mL). Red-violet coloration within one minute confirms the presence of aliphatic primary amine.

Note

- Aliphatic secondary amines will also give this test.

Safe Waste Disposal

- Remove the precipitate, if any, and transfer it to solid organic waste container. Place the residual liquid in the aqueous waste container.

(d) *Azo dye test:* This test is specific for primary aromatic amines. It is based on the fact that the primary aromatic amines form stable diazonium salts, at 0-5°C, which couple with phenols in the weakly basic pH to form orange-red coloured dye. The reaction is highly temperature and pH sensitive.

OH + N₂Cl → (dil. NaOH, pH 9-10) → 1-Phenylazo-2-naphthol or Sudan-1 (red-orange) + NaCl + H₂O

2-Naphthol + Phenyldiazonium chloride

Procedure: In case a clear solution is obtained in nitrous acid test, see Table 4.27, without evolution of nitrogen gas at room temperature, then add this ice-cold solution (2.0 mL) to a solution of 2-naphthol in 10% aqueous sodium hydroxide (4.0 mL), pre-cooled in ice. Appearance of orange-red coloured dye confirms the presence of primary aromatic amine.

Note

- Being unstable, diazonium salts of primary aliphatic amines decompose under the reaction conditions and liberate nitrogen gas.
- If the clear solution obtained in nitrous acid test is warmed, evolution of nitrogen gas takes place due to decomposition of arene diazonium salts at higher temperature.
- Arene diazonium salts, if allowed to dry, can lead to explosion. So, it is important that they are consumed immediately upon formation.
- For detailed theory, please see *section 6.5*, Volume 1.

Safe Waste Disposal

- Dilute the test solution with water (10.0 mL) and pass through charcoal (5.0 g). Place the charcoal in non-hazardous waste container. Place the filtrate in the aqueous waste container.

4. Confirmatory test for secondary amines (Liebermann Nitrosamine test)
Secondary aromatic or aliphatic amines react with nitrous acid, at low temperature, to form *N*-nitrosamine which precipitates as yellow-orange oil, due to decreased basicity. The *N*-nitrosamine generates nitrous acid on reaction with concentrated sulphuric acid. The nitrous acid then reacts with phenol to form yellow coloured 4-nitrosophenol (quinone monoxime), which further condenses with excess of phenol to form blue/green coloured indophenol cation. On dilution, indophenol exhibits red coloured solution which turns blue or green again on basification with sodium hydroxide solution, due to formation of indophenol anion.

$$R_2N\text{--}N{=}O \xrightarrow{\ H_3O^{\oplus}\ } R_2NH + HONO$$

N-Nitrosamine

Procedure: Extract the yellow oil obtained in nitrous acid test (see Table 4.27) by adding ether (5.0 mL), wash the organic layer successively with water, dilute NaOH and then again water and evaporate to dryness carefully. Take the *N*-nitrosamine residue (0.05 g) in a clean, dry test tube and add phenol (0.05 g) and warm for 20 seconds, cool and add concentrated sulphuric acid (2.0 mL). Intense blue/green colouration which turns red on dilution with water and changes again to blue/green on basification, confirms secondary amine.

Note

- This reaction is normally characteristic of phenols in which a para-position is unsubstituted. If *p*-position is occupied, the reaction may proceed to ortho-position.

- Washing ethereal layer with dilute NaOH neutralizes excess of unreacted nitrous acid which otherwise prevents coupling of *p*-nitrosophenol with phenol, by converting all the unreacted phenol to *p*-nitrosophenol.

- These compounds are carcinogenic therefore should be handled with care. Consult MSDS. (https://Pubchem.ncbi.nlm.mi.gov)

Safe Waste Disposal

- Excess of isolated nitrosamine should be placed in hazardous solid waste container. Make the test solution basic by adding 10% aqueous sodium hydroxide. Transfer the mixture to the aqueous solution container.

5. Confirmatory tests for tertiary aromatic amine: Tertiary aromatic amines can be confirmed by formation of orange coloured solution with nitrous acid (See Table 4.27). Basification of the solution with 10% aqueous sodium hydroxide forms a green solid due to *p*-nitrosamine. For details see nitrous acid test discussed earlier on page 205.

6. Spot tests for amines

(a) *Azodye test:* Apply the following spots on a TLC plate or Whatman paper strip sequentially, as described below:

- Dilute aqueous solution of the compound in water/alcohol
- Dilute HCl

- Aqueous sodium nitrite
- Alkaline solution of 2-naphthol

Development of orange-red spot confirms the presence of primary aromatic amine.

(b) *Liebermann nitrosamine test for secondary amines:* Spot the following solutions on a Merck strip, sequentially, as described below:

- Liquid compound or concentrated solution of solid in alcohol
- Aqueous sodium nitrite
- Concentrated sulphuric acid
- Heat the Merck strip on preheated oven (~120°C) for 1 minute and apply the spots of the following solutions on top of the same spot
- Saturated solution of urea (to remove excess of nitrous acid)
- Concentrated alcoholic solution of phenol
- Heat the Merck strip on preheated oven (~120°C) for 30 seconds and scratch the silica of the spotted portion and transfer to a beaker containing dilute sodium hydroxide solution and add 1-2 mL of water.

A blue coloured solution confirms the presence of secondary amine.

(c) *Nitrous acid test for tertiary amines:* Spot the following solutions on a Merck strip or Whatman paper, sequentially, as described below:

- Liquid compound or concentrated solution of solid in alcohol
- Concentrated hydrochloric acid
- Aqueous sodium nitrite

Appearance of orange spot indicates aromatic tertiary amine.

- Apply dilute sodium hydroxide solution 2-3 times.

The orange spot turns green confirming the presence of tertiary amine.

6. Spectroscopic identification

Table 4.28   Spectral Data for Amines

S. No.	Type of spectrum	Peak value	Peak attribution	Remarks
1.	IR	3400 & 3500 cm^{-1}	N-H Str. (free; symm. and antisymm.)	Doublet (primary amine)
		3350 cm^{-1}		Singlet (secondary amine)
		1580-1650 & 1515 cm^{-1}	N-H Str. (free) N-H bending	Primary and secondary amine
2.	NMR	0.5-3.0 δ	N-<u>H</u>*(aliphatic)	Broad**
		3.0-5.0 δ	N-<u>H</u>*(aromatic)	Broad**
		2.0-3.0 δ	>C<u>H</u>-N<	-

*Position varies with solvent and concentration; ** replaceable with D_2O

Derivatives of Amines

1. Benzoyl derivative: The N-H bond in primary and secondary amines provides a suitable route to derivatize these compounds as amides of acetic (acetyl derivative) or benzoic acid (benzoyl derivative). In the *Schotten-Baumann* method of benzoylation, primary and secondary amine is suspended in an excess of 10% aqueous sodium hydroxide solution and benzoyl chloride is added. On vigorous shaking, the solid benzoyl compound, being insoluble in water, separates out.

Tertiary amines do not show this reaction due to lack of replaceable hydrogen on the nitrogen atom (for detailed theory, see Volume 1, *section 6.8*).

$$RNH_2 \;+\; \underset{\substack{\text{Benzoyl}\\\text{chloride}}}{\underset{Ph\quad Cl}{\overset{O}{\|}}} \;\xrightarrow{\text{Aq. NaOH}}\; \underset{\text{Benzoyl derivatives}}{\underset{RHN\quad Ph}{\overset{O}{\|}}} \;+\; NaCl \;+\; H_2O$$

1° amine

$$RR'NH \;+\; \underset{\substack{\text{Benzoyl}\\\text{chloride}}}{\underset{Ph\quad Cl}{\overset{O}{\|}}} \;\xrightarrow{\text{Aq. NaOH}}\; \underset{\text{Benzoyl derivatives}}{\underset{R'RN\quad Ph}{\overset{O}{\|}}} \;+\; NaCl \;+\; H_2O$$

2° amine

Procedure: Place the given amine (1.0 mL) and 10% aqueous NaOH (15.0 mL) in a conical flask fitted with a stopper. Add first lot of benzoyl chloride (1.5 mL, added in two lots of 0.75 mL each) and stopper the flask immediately. Shake the reaction mixture vigorously for 10-15 minutes or till contents come back to room temperature. Add the second lot of benzoyl chloride and repeat as stated above. Filter the separated solid on suction pump, wash well with water. Dry the solid and recrystallize a fraction of the crude sample using ethanol. Filter the crystallized solid, dry on a porous clay plate and record its melting point.

Note

- In case a sticky solid is obtained on benzoylation, isolate the solid and allow it to stand at room temperature in ~5% aqueous NaOH, for 5-10 minutes. This will remove any unreacted benzoyl chloride.

- In case a sticky solid still persists after sodium hydroxide treatment, suspend the solid in 5% aqueous HCl, to remove any unreacted amine. Give a final washing with water before recrystallization.

Safe Waste Disposal

- Place the test solution in aqueous solution container after neutralizing with 5% aqueous HCl to pH ~7.

2. Acetyl derivative: Acetylation of primary and secondary amines leads to formation of secondary and tertiary amides ($RNHCOCH_3$ or R_2NCOCH_3) respectively. Acetylated compounds are generally crystalline substances and can be used for identification of amines. Tertiary amines do not form the acetyl derivative since they lack replaceable hydrogen on the nitrogen atom. Most amines are water insoluble, therefore they are first dissolved in dilute acid and then treated with acetic anhydride and sodium acetate. Sodium acetate liberates the free base from the salt, which then reacts with acetic anhydride to form the acetyl derivative.

In green method, acetylation of primary amines can be directly carried out with acetic acid in the presence of zinc dust. The zinc acetate formed during the reaction actually acts as the acetylating agent. (For detailed theory, see Volume I, *section 6.7*)

$$RNH_2 \; + \; 2CH_3COOH \xrightarrow[\text{Reflux}]{\text{Zn dust}} \underset{RHN \quad CH_3}{\overset{O}{\underset{\|}{}}} + \; ZnO \; + \; CH_3COOH \; + H_2$$

Amine   **Gl. Acetic acid**   **Acetyl derivatives**

$$RNH_2 \; + \; (CH_3CO)_2O \xrightarrow[\text{CH}_3\text{COONa}]{\text{HCl}} \underset{RHN \quad CH_3}{\overset{O}{\underset{\|}{}}} + \; 2CH_3COOH \; + \; NaCl$$

Amine   **Acetic anhydride**   **Acetyl derivatives**

Procedure: Place the given primary amine (1.0 mL), glacial acetic acid (3.0 mL), Zn dust (0.05 g) along with some pumice stones in a round bottomed flask, fitted with a water condenser. Reflux the contents gently over wire guaze for 90 minutes. Filter the contents hot to remove any unreacted zinc. Pour the contents in a beaker containing crushed ice and stir with glass rod. Filter the separated solid on suction pump, wash well with water. Dry the solid and recrystallize a fraction of the crude sample using hot water or aqueous ethanol. Filter the crystallized solid, dry on a porous clay plate and record its melting point.

Alternatively, in a clean conical flask (250 mL), dissolve freshly distilled amine (1.0 mL) in dilute hydrochloric acid (4-5 mL). Cool the contents in ice bath. To this cooled solution, add acetic anhydride (3.0 mL), followed by sodium acetate solution [prepared by dissolving sodium acetate trihydrate (2.5 g) in water (2.5 mL)]. Cool the flask in ice-bath and stir the contents with glass rod for 10-15 min. Filter the separated solid on suction pump, wash well with water. Dry the solid and recrystallize a small fraction of crude sample from hot water/aqueous ethanol. Filter the crystallized solid, dry on a porous clay plate and record its melting point.

Note

- Ensure that the smell of acetic anhydride has disappeared completely, before filtering the crude solid.
- Acetyl derivatives of certain amines are water soluble therefore avoid adding exces of water.
- Diacetyl derivatives of primary amines formed during the reaction are hydrolysed to monoacetyl derivative, upon recrystallization from aqueous ethanol.
- If the amine is water soluble then the following modified procedure can be used for acetylation: Dissolve the given amine (0.5 g or 1.0 mL) in ice-cold water (10.0 mL) and add acetic anhydride (3.0 mL). Shake for 10-15 minutes. Filter the precipitated solid then proceed as stated above.

Safe Waste Disposal

- Place all the test solutions in the aqueous waste container after neutralization to pH ~7 with 5% aqueous NaOH or HCl (as applicable).

3. Bromo derivative: Primary aromatic amines undergo electrophilic substitution reaction, at o-/p-positions, to produce corresponding bromo derivatives. (for detailed theory, see Volume 1, *section 6.3*, Bromination)

$$\text{Aniline} \; (NH_2\text{-C}_6H_5) \; + \; 3Br_2 \; \xrightarrow{\text{Glacial acetic acid}} \; \text{2,4,6-Tribromoaniline} \; + \; 3HBr$$

Procedure: Add solution of bromine (1.6 mL) in glacial acetic acid (4.0 mL) dropwise (from burette), to a solution of freshly distilled aniline (1.0 mL) in glacial acetic acid (4.0 mL), placed in a 100 mL conical flask. Cool the contents in an ice-bath, during addition, and shake the reaction mixture occasionally. After addition is complete, cover the conical flask with filter paper. Allow the reaction mixture to stand for 10-15 minutes. Completion of reaction is indicated by the appearance of faint orange-yellow colour due to slight excess of bromine. Transfer the contents into a beaker containing ice-cold water (~50 mL). A white coloured solid is obtained, filter the separated solid on suction pump, wash well with water. Dry the solid and recrystallize a part of the crude sample using aqueous ethanol. Filter the crystallized solid, dry on a porous clay plate and record its melting point.

Note

- Bromination should be attempted only for those amines where acetylation or benzoylation fails.
- Bromine is highly corrosive, toxic and is a severe environmental pollutant. Its use should be avoided as far as possible.

Safe Waste Disposal

- Place all the test solutions in halogenated organic waste container.

4. Picrate: Like polynuclear hydrocarbons, tertiary amines undergo reaction with picric acid to give addition products (π-complexes) with definite melting point. Picrates of tertiary amines are more stable than those of polynuclear hydrocarbons. (For detailed theory, see derivatives of aromatic hydrocarbons, see *section 4.3.2.4,* page 198)

$$
\underset{\textbf{3° Amine}}{R\!-\!N\!\overset{R}{\underset{R}{<}}} \;+\; \underset{\textbf{Picric acid}}{\text{O}_2\text{N}\!-\!\overset{\text{OH}}{\underset{\text{NO}_2}{\bigcirc}}\!-\!\text{NO}_2} \longrightarrow \underset{\textbf{Picrate}}{R\!-\!\overset{R}{\underset{R}{N}}\!\cdot\!\text{O}_2\text{N}\!-\!\overset{\text{OH}}{\underset{\text{NO}_2}{\bigcirc}}\!-\!\text{NO}_2}
$$

Procedure: Dissolve the given tertiary amine (0.5 mL) in ethanol (5.0 mL) and add cold saturated solution of picric acid in ethanol (5.0 mL). Warm the contents on steam water bath and then transfer into a china dish. Allow the mixture to cool slowly. Filter the precipitated yellow solid, dry and recyrstallize a fraction of the crude sample using dilute ethanol. Filter the crystallized solid, dry on a porous clay plate and record its melting point.

Note

- If the compound does not dissolve in cold ethanol, warm the solution, cool and filter to get clear solution of amine.
- Some picrates are pure and do not require recrystallization.
- Do not allow picric acid to dry, as it may explode.
- Weakly basic amines like triphenyl amine and diphenyl amine do not form picrates.
- Picrates of primary and secondary amines can be prepared in the similar manner. Some of these picrates are stable. Check stability before attempting the preparation.

Safe Waste Disposal

- Place all the test solutions in the aqueous solution container.

Post Lab Questions

1. Why carbylamines test is usually not preferred in the lab?

2. Can you test aliphatic primary amines by Azo dye test? If not, why?

3. Azo dye test for primary aromatic amines must follow the following conditions:

   (a) Low temperature (0-5°C)

   (b) Mildly acidic pH (4-5)

   Explain what happens, if the medium is strongly acidic?

4. Sometimes amines give positive ferric chloride test. Give reasons.

5. During Azo dye test, sometimes a yellow precipitate separates on addition of aqueous sodium nitrite to solution of amine. Give reason and explain how you can avoid it.

6. Which of the following is more basic and why:

   (a) N-Methylamine and N, N-dimethylamine

   (b) Triethyl amine and ethylamine

   (c) Methylamine and aniline

   (d) Aniline, o-toluidine, 2,6-xylidine

7. What happens when:

   (a) Concentrated HCl is added to aniline

   (b) 2,4-DNP reagent is added to aniline solution

8. Aniline cannot be subjected to nitration directly. Explain and give the structure of the major product formed.

9. How will you distinguish between aniline, N-methylaniline, N,N-Dimethyl-aniline using IR spectroscopy?

10. 4-(N-Methylamino) benzoic acid is subjected to Hinsberg's test. What do you think will happen after shaking with Hinsberg's reagent?

    (a) A white precipitate is formed

    (b) A clear solution is obtained.

    Give reason for your choice.

11. Explain the role of glacial acetic acid in the bromination of aniline.

12. What is the nitrosating agent in Liebermann nitrosamine reaction?

13. Secondary and tertiary amines do not give Rimini's test. Explain.

14. Distinguish between a primary aliphatic and primary aromatic amine by:

    (a) Physical method

    (b) Chemical method

4.3.2.6 *Compounds Containing Carbon, Hydrogen, Nitrogen and Oxygen (Neutral Compounds)*

Neutral organic compounds containing nitrogen are nitro, amides, anilides, imides and nitriles. Since most amino acids are neutral, barring a few acidic and basic amino acids, they have also been considered in this section along with other neutral nitrogen containing organic compounds. Most of these compounds are insoluble in dilute HCl, being neutral, except those which are water soluble.

4.3.2.6.1 *Nitro compounds*

They are compounds having general formulae (RNO_2 or $ArNO_2$). These compounds are generally insoluble in water and have yellow to deep yellow colouration due to the presence of chromophoric nitro group ($-NO_2$). For example, nitrobenzene is pale yellow liquid and settles at bottom, being heavier than water. It has a distinct bitter almond smell like benzaldehyde and benzonitrile. *p*-Nitrotoluene is a pale yellow solid whereas *m*-dinitrobenzene is a colourless solid when pure, but often acquires pale yellow colour on keeping. Nitro compounds can be reduced to distinct products in neutral/acidic/alkaline medium. These reactions can therefore be used for qualitative analysis of nitro compounds in the lab. The commonly used classification tests for nitro compounds are listed in Table 4.29.

Table 4.29 Classification Tests for Nitro Compounds

S. No.	Classification test	Test category	Observation	Inference
1.	Mulliken Barker test[a]	Group test	Black or grey residue.	Nitro compound
2.	Reduction with Sn/HCl followed by azo dye test	Special test	Orange-red dye	Aromatic nitro compound
3.	Reduction with ferrous hydroxide	Special test	Red brown/brown precipitate	Nitro compound
4.	Alcoholic sodium hydroxide test	Special test	Red or blue-violet precipitate	Trinitro/dinitro compound

a: To be performed for neutral compounds containing nitrogen as extra element

Classification Tests for Nitro compounds

1. Mulliken Barker test: The reaction involves reduction of nitro compound in neutral medium with zinc, in presence of aqueous ammonium chloride or calcium chloride, to form hydroxylamine. The hydroxylamine reduces Tollens' reagent to grey or black precipitate of metallic silver and itself gets oxidised to corresponding nitroso compound. This test is given by only aromatic nitro/ tertiary aliphatic nitro compounds.

$$RNO_2 + 4[H] \xrightarrow[NH_4Cl]{Zn} RNHOH + H_2O$$

Nitro compound · **Hydroxylamine**

$$RNHOH + 2Ag(NH_3)_2OH \longrightarrow RNO + 2H_2O + 2Ag\downarrow + 4NH_3$$

Tollens' Reagent

Procedure: Dissolve the unknown compound (0.2 mL or 0.2 g) in 50% ethanol (4.0 mL) taken in a clean test tube, add ammonium chloride (0.2 g) and zinc dust (0.2 g). Shake the contents and carefully boil for about five minutes on a low Bunsen burner flame. Filter the solution hot to remove excess of zinc dust. Transfer the filtrate (2-3 drops) to a clean test tube, add freshly prepared Tollens' reagent (2.0 mL) and look for any grey/black precipitate or silver mirror. If no change takes place in cold, heat the contents on the hot water bath for 2-3 minutes. Any grey/black precipitate or silver mirror on the inner walls of the test tube confirms the presence of nitro group in the compound.

Note

- Sometimes a grey precipitate is formed which on standing for 5-10 minutes turns black.

- A blank test of the unknown compound and Tollens' reagent should be done. The readers should note that nitro compounds do not reduce the Tollens' reagent whereas aldehydes and carbohydrates do. Some methyl ketones also reduce Tollens' reagent. Therefore, this test should not be done for compounds which contain aldehyde/methyl ketone group.

- Azoxy, nitroso and azo compounds are also reduced with zinc dust and ammonium chloride to products which reduce the Tollens' reagent.

- The filtrate (2-3 drops) may also be warmed with Fehling solution (2.0 mL) on a water bath, hydroxylamine will reduce cupric ions to produce red precipitate of cuprous oxide.

- Metallic silver produced should be immediately decomposed with dilute nitric acid to avoid forming explosive silver fulminate (AgCNO).

Safe Waste Disposal

- Transfer the test solution into a beaker. Add 5% aqueous nitric acid dropwise until metallic silver dissolves completely. Then neutralize the test solution by adding solid sodium carbonate in small lots (caution: foaming!). Next, add saturated sodium chloride solution (2.0 mL) to precipitate silver as silver chloride. Filter the solution and transfer silver chloride residue to the non-hazardous solid waste container. Transfer the filtrate to the aqueous waste solution container.

2. Reduction with tin and hydrochloric acid followed by Azo dye test: This reaction is given by aromatic nitro compounds only. It involves reduction of nitro compound in the acidic medium with tin and hydrochloric acid to form the corresponding primary amine which is then subjected to Azo dye test to yield orange-red coloured dye.

$$\text{Nitrobenzene} + 6[H] \xrightarrow[\text{HCl}]{\text{Sn}} \text{Aniline} + 2H_2O$$

Nitrobenzene

Aniline

$NaNO_2 + HCl$
(0-5°C)

1-Phenylazo-2-naphthol
or
Sudan-1

2-Naphthol
Dil. NaOH
(0-5°C)

Procedure: In a clean and dry round bottomed flask fitted with reflux condenser, add the given nitro compound (2-3 drops), tin granules (2-3 pieces), alcohol (2-3 mL) and then add concentrated hydrochloric acid (5.0 mL, in 2-3 lots, shake after every addition). Allow the vigorous reaction to subside, add 1-2 pumice stones and connect the water condenser. Reflux the contents on the boiling water bath for 20-30 minutes. Cool the reaction mixture by adding ice, neutralize with 20-40% aqueous sodium hydroxide until all the precipitate of tin hydroxide dissolves. Extract with ether (5.0 mL) and evaporate ether carefully. Perform the azo dye test with the residue. (See **Azo dye test**, under classification tests for amines, *section 4.3.2.5*, page 209)

Note

- Do not perform this test if the compound already contains an aromatic amino group.
- Perform blank Azo dye test with parent compound before you draw any conclusion.
- Ensure that the oily droplets due to nitroarenes have disappeared before work up.
- Handle ether carefully as it is flammable.

Safe Waste Disposal

Neutralize the aqueous filtrate with 10% aqueous hydrochloric acid, separate the tin hydroxide by filtration and place in a non-hazardous solid waste container. Place the aqueous filtrate in aqueous solution container. Place the recovered ether in the organic solvent container.

3. Reduction with ferrous hydroxide: Nitro compounds can be reduced in the alkaline conditions by ferrous hydroxide to corresponding amines. Ferrous hydroxide required for the reaction is generated from Mohr's salt and potassium hydroxide. A positive test is indicated by the change in colour from green to red-brown or brown due to the oxidation of ferrous ions to ferric ions.

$$RNO_2 + 6Fe(OH)_2 + 4H_2O \longrightarrow RNH_2 + 6Fe(OH)_3$$

Nitro compound / Ferrous hydroxide / 1° amine / Ferric hydroxide (red-brown/brown)

Procedure: Take the ferrous sulphate reagent (1.0 mL, See Appendix II) in a small eppendorf (2.0 mL), add the nitro compound (0.01 g or 2-3 drops, if liquid) followed by addition of 3% alcoholic potassium hydroxide reagent (0.7 mL). Stopper quickly, and shake for 1 minute. Immediate formation of red-brown to brown precipitate of iron(III) hydroxide confirms the presence of nitro compound.

Note

- The speed with which the nitro compound is reduced depends on its solubility. For example, *p*-nitrobenzoic acid, which is soluble in the basic reagent, will give a positive test instantly.
- Compounds like quinones, nitroso compounds, hydroxylamines, alkyl nitrites and alkyl nitrates, which oxidise ferrous hydroxide, will also give a positive test.
- This reaction cannot be used for testing highly coloured compounds.
- Perform test in the eppendorf and fill the solution to the brim. This will push out any air and prevent aerial oxidation of ferrous ions.
- Green colour indicates negative test.

Safe Waste Disposal

- Filter the ferric hydroxide and transfer the residue to the non-hazardous solid waste container. Add 10% aqueous hydrochloric acid to the filtrate until it is neutral to pH paper and then transfer the solution to aqueous waste solution container.

4. Reaction with alcoholic sodium hydroxide: This is a colour test which can be used to distinguish between aromatic compounds containing mono/di/tri-nitro groups from each other based on the coloured complex they form with alcoholic sodium hydroxide. The colouration is due to the formation of *Meisenheimer complex* of the nitro compounds.

Meisenheimer complex

Procedure: Take 20% sodium hydroxide solution (5.0 mL) in a clean test tube and add ethanol (2.0 mL) followed by the nitro compound (1 drop or a crystal). Shake the mixture vigorously. Note the colour of the solution. Mononitro compounds do not show change in colour, dinitro compounds produce blue-purple colour whereas trinitro compounds produce red colouration.

Note

- This test should be performed if Mulliken Barker and ferrous hydroxide tests are positive.

- The presence of a hydroxyl, amino, substituted amino group in the molecule prevents formation of the characteristic red/purple colouration. For example, *p*-nitrophenol gives yellow colouration due to resonance stabilized phenoxide ion:

- Mononitro hydrocarbons are not strongly electrophilic to facilitate nucleophilic attack on the ring. However, nitrohalides may show activated nucleophilic substitution.

Safe Waste Disposal

- Transfer the test solution to the aqueous waste solution container.

5. Spot tests

(a) *Reduction with Zinc/HCl followed by Azodye test:*   Spot 1-2 drops of the following solutions on a Whatman paper, sequentially, as described below:

- Concentrated alcoholic solution of the compound
- Dilute hydrochloric acid.
- Rub gently zinc dust (5.0 mg) on the same spot using ear bud and wait for 30 seconds. Check original paper for heating. Heat the paper strip in the oven maintained ~120°C for 20-30 sec. and then take out the strip.
- Reverse the paper and again spot dilute hydrochoric acid on the spotted portion.
- Introduce sodium nitrite solution (1-2 drops) from the kit.
- Add dilute alkaline solution of 2-naphthol (1-2 drops)

Development of orange-red colour confirms the presence of aromatic nitro compound.

Note

- Perform simultaneously a blank test (apply all the solutions except Zn dust).
- The test can be performed only with those compounds which do not contain a primary aromatic amino group.

6. Spectroscopic identification

Table 4.30   Spectral Data of Nitro-hydrocarbons

S. No.	Type of spectrum	Peak value	Peak attribution	Remarks
1.	IR	1290-1390 & 1500-1550 cm^{-1}	-N=O Str.(Sym. and antisym.) [Aromatic compounds]	Doublet
		1372 & 1550 cm^{-1}	-N=O Str.(Sym. and antisym.) [Aliphatic compounds]	
2.	NMR	4.1-4.4 δ	>C$\underline{H}$NO$_2$	
		8.1 δ (*o*-protons)		
		7.5 δ (*m*-protons)	ArNO$_2$	
		7.65 δ (*p*-protons)		

Derivatives of Nitro Compounds

1. Nitro derivative: Almost all mononitro compounds can be nitrated to polynitro compounds. For example, nitrobenzene on nitration with nitrating mixture yields *m*-dinitrobenzene. (For detailed theory, see Volume I, *section 6.4*)

$$\text{Nitrobenzene} + \text{conc. HNO}_3 \xrightarrow{\text{conc. H}_2\text{SO}_4} \text{m-Dinitrobenzene} + \text{H}_2\text{O}$$

Procedure: In a 100 mL dry round bottomed flask, prepare nitrating mixture by carefully adding concentrated sulphuric acid (2.0 mL) dropwise with shaking to fuming nitric acid (1.5 mL), cool in an ice-water bath. Add 1-2 pumice stones and then add dropwise the given nitro compound (1.0 mL), with occasional stirring. Remove the flask from the ice bath, put 1-2 pumice stones, attach a water condenser and reflux the contents on a boiling water bath for 30-45 minutes. Thereafter, pour the contents into a 100 mL beaker containing crushed ice and stir with the help a glass rod. Filter the precipitated yellow solid at the suction pump. Wash with water (to remove excess acid) until the filtrate is colourless. Dry and recrystallize a fraction of crude sample with ethanol. Filter, dry the purified product on porous plate. Report the melting point of recrystallized sample.

Note

- See nitration of halohydrocarbon, *section 4.3.2.3*, page 194.

Safe Waste Disposal

- See nitration of aromatic hydrocarbons, *section 4.3.2.4*, page 199.

2. Reduction to amine: Depending upon the pH of the reaction medium and the type of reducing agent used, aromatic nitro compounds are reduced to different products. The reduction reaction may be carried out in the acidic (Sn/HCl)/alkaline (Zn/NaOH)/neutral medium (Zn/NH$_4$Cl). The products obtained under each of these conditions will be different. Iron being cheaper than tin is used in place of tin for commercial preparation of aniline. The aromatic amine so formed can be identified by convertion to acetyl or benzoyl derivative. (See acetylation/benzoylation under derivatives of amines, *section 4.3.2.5.1*, page 213-214)

$$\text{Nitrobenzene} + 6[H] \xrightarrow{\text{Sn/HCl}} \text{Aniline} + 2H_2O$$

Procedure: Take the given nitro compound (1.0 mL) and granulated tin (2.0 g) in a clean and dry round bottomed flask. Attach a reflux condenser. Add concentrated hydrochloric acid (10.0 mL), in small lots, with constant shaking. Cool the contents in a ice-water bath, if the contents become hot. Allow the reaction mixture to stand for 10 minutes at room temperature. If the mixture is turbid, add ethanol (5.0 mL). Add 1-2 pumice stones and reflux the contents on a wire gauze for 30-60 minutes. Take a drop of reaction mixture in a test tube and add ice-water (5.0 mL), appearance of clear solution indicates that the reaction is complete. Transfer the contents to a beaker (250 mL) and allow to cool. Neutralize the solution by adding 20-30% aqueous sodium hydroxide until precipitate of tin hydroxide dissolves completely. Filter, if a solid is precipitated, recrystallize and determine its melting point. Else extract the amine with two portions of ether (2 × 10 mL). Separate the ether layer, wash well with water and dry over anhydrous potassium carbonate. Filter potassium carbonate and distill ether carefully in a rotatory evaporator. Use the residual amine to form acetyl/benzoyl derivative.

Alternatively, place the given nitro compound (0.5 g), zinc dust (2.5 g) and alcohol (4-5 mL) in a clean and dry conical flask. Add hydrochloric acid (15.0 mL) in small lots with constant shaking. The completion of reaction is indicated by the complete dissolution of the initially insoluble nitro compound. Add ice (50 g) to cool the reaction mixture. Neutralize with 20-30% aqueous sodium hydroxide solution. Then proceed as described above.

Note

- Concentrated hydrochloric acid should be added in lots of 1 mL each. After every addition, the reaction mixture heats up and starts to boil. Allow initial reaction to subside before adding the next lot.

- Before workup, ensure no oily droplets remain in the reaction mixture and the mixture does not smell of nitrobenzene.

- A clear solution indicates that that the original nitro compound has been reduced to amine which dissolves in the acid due to salt formation.

- Double the amounts of reagent may be used for reduction of dinitro compounds. If the reduced amine is a crystalline solid, it may identified by recrystallization followed by Azo dye test and melting point determination.

- While neutralization, cool the reaction mixtre in a ice-water bath. This avoids vigorous reaction.

- The alternate method works well for mononitro compounds only. This method is energy efficient as no heating is required.

Safe Waste Disposal

- See tin/HCl reduction under classification tests for nitro compounds, *section 4.3.2.6.1, page 219.*

3. Selective reduction of polynitro compounds: In aromatic compounds containing two or more nitro groups, it is possible to reduce one of those groups by using suitable reducing agents. This type of reduction is called *selective reduction.* Zinin reduction involves reduction of polynitro compounds with sodium or ammonium sulphide, hydrosulpide or polysulphides in the presence of excess of water, to form nitroamines. For example, the reduction of *m*-dinitrobenzene with sodium disulphide gives *m*-nitroaniline.

NO$_2$ / NO$_2$ (*m*-Dinitrobenzene) $+$ Na$_2$S$_2$ $\xrightarrow[\text{(Excess)}]{\text{H}_2\text{O}}$ NO$_2$ / NH$_2$ (*m*-Nitroaniline) $+$ Na$_2$S$_2$O$_3$

For detailed theory, see Volume 1, *section 6.11.7.*

Procedure: Take finely powdered sulphur (0.4 g) and crystalline sodium sulphide (1.6 g) in a beaker and add water (6.0 mL). Boil the mixture gently on a wire gauze until contents get dissolved to give deep brown coloured solution of sodium disulphide.

On a sand bath, heat *m*-dinitrobenzene (1.0 g) with water (25.0 mL), until the water boils. Remove the mixture from the sand bath. Add hot sodium disulphide solution dropwise (over a period of ~30 minutes) to the boiling aqueous mixture of *m*-dinitrobenzene, with constant stirring. After complete addition, heat the solution gently on the sand bath for the 20 minutes and then vigorously for the next five minutes. Filter hot through a preheated funnel and conical flask, using a thin cotton plug. Pale yellow crystals of *m*-nitroaniline will separate out on cooling the clear filtrate. Filter the crystals at suction, wash well with minimum amount of cold water, and dry. Recrystallize a fraction of the crude sample from hot water/aqueous ethanol. Filter the crystallized solid, dry on a porous clay plate and record its melting point.

Note

- Do not allow the volume of the reaction mixture to fall below 25 mL anytime during the reaction.

- Addition of hot solution of sodium disulphide should be done slowly, dropwise with constant stirring.

Post Lab Questions

1. Write the product of reduction of nitrobenzene with each of the following:
   (a) Sn/HCl
   (b) Zn/HCl
   (c) Zn/NaOH/MeOH
   (d) Zn/NaOH

2. What happens when *m*-dinitrobenzene reacts with:
   (a) Sn/HCl
   (b) Na_2S/S
   (c) Na_2S/S in excess

3. Give one test to distinguish between nitromethane and nitrobenzene.

4. What happens when *p*-nitrophenol and *m*-dinitrobenzene are individually subjected to reaction with alcoholic sodium hydroxide?

4.3.2.6.2   *Amides, imides and nitriles*

Amides are derivatives of carboxylic acids where the –OH group in –COOH is replaced by NH_2 group. Imides are cyclic amides of dicarboxylic acids whereas nitriles are carboxylic acid derivatives having general formula, RCN. Though these are different compounds they can be classified by some common qualitative tests described in Table 4.31.

Amides are colourless, odourless solids. However formamide, $HCONH_2$, is a liquid and acetamide has a mouse-like odour unless pure. Lower aliphatic amides like formamide, acetamide, and urea are readily soluble in cold water. Aromatic amides like benzamide and salicylamide are almost insoluble in cold water but dissolve in hot water. Salicylamide is readily soluble in cold NaOH solution due to the presence of –OH group.

Imides are normally colourless solids. Aliphatic imides like succinimide is readily soluble whereas aromatic imides like phthalimide are sparingly soluble in cold water.

Lower aliphatic nitriles like acetonitrile is miscible with water but aromatic compounds like benzonitrile are practically insoluble. Acetonitrile maintains the mouse-like odour of acetamide unless purified. Benzonitrile has a bitter almond smell like benzaldehyde and nitrobenzene. Classification tests for amides, imides and nitriles are described in this section. However, detailed treatment is taken up only for amides, as other two class of compounds are not dealt at the UG level labs.

Table 4.31 Classification Tests for Amides, Imides and Nitriles

S. No.	Test	Category	Observation	Inference
1.	Hydrolysis with Alkali	Group test	Smell of ammonia	Amide/imide/Nitrile
2.	Hydroxamic acid test	Special test	Wine-red or violet colouration	Amide or Nitrile[a]
3.	Reaction with KOH	Special test	White precipitate	Imide
4.	Biuret test	Special test	Violet colouration	Urea

a: This test is also given by esters but they do not liberate ammonia in alkaline hydrolysis.

Classification tests for amides, imides and nitriles

1. Hydrolysis with alkali: Amides, imides and nitriles are hydrolysed by alkali to liberate ammonia and form soluble sodium salt of carboxylic acids. Acidification of alkaline solution precipitates the carboxylic acids whereas ammonia is detected by its unpleasant odour or with thick white fumes of ammonium chloride, when a rod dipped in concentrated HCl is brought to the mouth of the reflux condenser.

$$RCONH_2 \ + \ NaOH \ \xrightarrow{\ H_2O\ } \ RCOONa \ + \ NH_3$$
Amide
R=Alkyl/Aryl

$$\text{Succinimide} \ + \ 2\,NaOH \ \xrightarrow{\ H_2O\ } \ \text{(diCOONa)} \ + \ NH_3$$
Succinimide

$$RCN \ + \ NaOH \ \xrightarrow{\ H_2O\ } \ RCOONa \ + \ NH_3$$
Nitrile
R=Alkyl/Aryl

Procedure: Take the unknown compound (0.2 g) in a round bottomed flask fitted with a reflux condenser and add 10% aqueous sodium hydroxide solution (2-3 mL). Mix the contents well. Add 1-2 pumice stone and reflux the solution gently on a small Bunsen burner flame for 3-4 minutes and note the odour. Bring a rod dipped in concentrated hydrochloric acid to the mouth of the condenser. Look for thick white fumes of ammonium chloride. Alternatively, bring red litmus paper to the mouth of the condenser. Evolution of ammonia will turn the red litmus paper to blue, else ammonia will turn the filter paper moistened with copper sulphate solution to blue. Cool the flask in tap water and acidify the solution with excess of dilute sulphuric acid. A colourless precipitate will separate out on acidification, if the carboxylic acid formed on hydrolysis is not water soluble.

Note

- Do not smell the liberated ammonia directly.
- Appearance of white precipitate indicates positive test for aromatic amide/imide/nitrile provided esters have already been ruled out.
- In case the carboxylic acid formed is cold water soluble, no precipitation is observed. In such case evolution of ammonia is an indication of the presence of amide/imide/nitrile.
- Only primary amides, imides and nitriles will liberate ammonia and not secondary or tertiary.
- Evolution of brisk effervescence of CO_2 with no precipitate indicates urea.
- Nitriles are best hydrolysed by boiling with 30-40% aqueous sodium hydroxide since the reaction is less rapid than primary amides.

Safe Waste Disposal

- Transfer the test solution to the aqueous waste solution container after neutralization to pH ~7 with 5% aqueous NaOH.

2. Hydroxamic acid test: Amides and nitriles combine with hydroxylamine to form hydroxamic acid. The hydroxamic acid further reacts with ferric chloride in the acidic medium to give burgundy or violet colouration due to formation of ferric hydroxamate.

$$RCONH_2 + NH_2OH.HCl \xrightarrow{\text{NaOH}} RCONHOH + NaCl + H_2O + NH_3$$

Amide   Hydroxylamine   Hydroxamic acid
        hydrochloride

$$RCN + NH_2OH \longrightarrow R-\overset{\overset{\displaystyle NH}{\|}}{C}-NHOH \xrightarrow[H^{\oplus}]{H_2O} R-\overset{\overset{\displaystyle O}{\|}}{C}-NHOH + NH_4^{\oplus}$$

$$3RCONHOH + FeCl_3 \longrightarrow \left[\begin{array}{c} R-C=O \\ | \\ HN-O \end{array} \right]_3 Fe + 3\ HCl$$

Ferric hydroxamate

Procedure: Take a solution of the unknown compound [0.05 g or 1-2 drops in propylene glycol (2-3 mL)] in a clean and dry test tube. Add a solution of hydroxylamine hydrochloride in propylene glycol (2.0 mL, 1M), followed by addition of aqueous potassium hydroxide (2.0 mL, 1M). Heat the reaction mixture for 2 minutes, cool and acidify with dilute hydrochloric acid. Add 5% alcoholic ferric chloride solution (1-2 drops). Appearance of burgundy red or violet colour confirms the presence of amide or nitrile.

Note

- Primary amides and nitriles will give this test.

- This test is also given by esters, acid chlorides and acid anhydrides. Therefore, this test may be taken as a positive test for amide or nitrile if these functional groups are absent.

- For aromatic primary amides proceed as follows: Place the unknown (0.05 g in 5.0 mL water) in a clean, dry test tube. Add 3% hydrogen peroxide (0.5 mL) and two drops of 5% aqueous ferric chloride solution. Heat the contents to boiling. The characteristic violet colour appears in case primary aromatic amide is present. Hydrogen peroxide is used to convert aromatic amide to hydroxamic acid.

$$\underset{\text{Aromatic amide}}{ArCONH_2} + H_2O_2 \longrightarrow \underset{\text{Hydroxamic acid}}{ArCONHOH} + H_2O$$

Safe Waste Disposal

- Transfer the acidic reaction mixture to a small beaker (100 mL), add solid sodium carbonate until neutral to pH paper (caution: foaming) then transfer the contents to the aqueous waste solution container.

3. Potassium hydroxide test for imides: Imides produce a white precipitate of potassium imide when reacted with potassium hydroxide. This is due to the presence of acidic hydrogen. This test will not be given by amides or nitriles.

Succinimide + KOH ⟶ Potassium succinimide (white ppt.) + H₂O

Procedure: Take saturated alcoholic solution of potassium hydroxide (1.0 mL) in a test tube and add saturated solution of the unknown compound in alcohol (1.0 mL) to it. Formation of white precipitate confirms the presence of imide.

Safe Waste Disposal

- Neutralize the filtrate with 5% aqueous hydrochloric acid to pH ~7. Transfer to aqueous solution container.

4. Biuret test for urea: Urea is an amide of carbamic acid, which on heating under dry conditions undergoes condensation to form biuret, with elimination of a molecule of ammonia. Biuret complexes with copper sulphate in alkaline solution to yield a blue-violet coloured solution.

Procedure: Take the given compound (0.05 g) in a clean and dry china dish and heat gently on a wire gauze. The compound first melts and then solidifies eliminating the ammonia gas. Remove the china dish from the flame, cool and add a drop or two of dilute sodium hydroxide solution. Then add 1-2 drops of very dilute copper sulphate solution. Appearance of blue-violet colouration confirms the presence of urea.

Note

- Heat the compound gently otherwise it will get charred.
- Do not add excess of sodium hydroxide as copper ions will get precipitated as copper hydroxide and will no longer react with biuret.
- Do not add excess of copper sulphate as the blue colour of the reagent will mask the ensuing blue-violet colour.
- Compounds like proteins, oxamide and malonamide also give this test.

5. Spot tests for aromatic amides:

(a) *Hoffmann bromamide reaction followed by Azo dye test:* Prepare concentrated aqueous or alcoholic solution of the compound. Spot 1-2 drops of the following solutions on a TLC plate or Merck or Whatman paper strip, sequentially, as described below:

- Concentrated solution of the compound
- Dilute NaOH solution
- Bromine solution in CCl_4
- Heat the strip in oven maitained at 120°C for 20-30 seconds
- The amide is converted into an aromatic primary amine having one carbon less

Remove the strip from oven and spot the following solutions, sequentially, on the same spot:

- Concentrated hydrochloric acid
- Aqueous solution of sodium nitrite
- Alkaline solution of 2-naphthol

Development of orange-red dye confirms the presence of primary aromatic amide.

Note

- A blank test should be simultaneously performed.
- This test is applicable to compounds lacking primary aromatic amino group.
- Sometimes an orange spot appears before addition of 2-naphthol due to liberation of bromine from sodium hypobromide. But this spot disappears on standing.

6. Spectroscopic identification

Table 4.32 Spectral Data for Amides

S. No.	Type of spectrum	Peak value	Peak attribution	Remarks
1.	IR	$3180\ \&\ 3350\ cm^{-1}$	N-H Str. (Solid sample) Primary amides	Doublet
		$1646\text{-}1695\ cm^{-1}$	>C=O Str. (Solid sample)	
		$1620\text{-}1655\ cm^{-1}$	N-H bending (Primary) Solid sample	
		$1515\text{-}1570\ cm^{-1}$	N-H bending (Secondary)	
2.	IR	$1700\text{-}1800\ cm^{-1}$	>C=O Str (Imides)	Doublet
		$3200\ cm^{-1}$	N-H Str.	Singlet
		$2250\ cm^{-1}$	$-C\equiv N$ Str. (Nitriles)	Weak band
3.	NMR	5-9 δ	-CONH$_2$[a]	replaceable by D_2O for imides.
		2-3 δ	>CHCONH$_2$/ >CHCN	-

a: position varies with concentration, temperature and solvent

Derivatives of amides

1. Alkaline hydrolysis: Hydrolysis of amides, which involves replacement of $C\text{-}NH_2$ linkage with C-OH, is extremely slow with water alone. However, it can be accelerated by carrying out the reaction either in alkaline or acidic conditions. Under alkaline conditions, salt of the acid and ammonia are formed whereas under acidic conditions ammonium salt and carboxylic acid are formed. The choice of acid or alkali as catalyst for hydrolysis depends upon the class of compounds, its solubility in the medium and effect of reagent on the hydrolytic product. Primary and low molecular weight substituted amides may be hydrolyzed by boiling with 10% aqueous sodium hydroxide or 10% aqueous sulphuric acid.

$$\underset{\text{Amide}}{RCONH_2} + NaOH \xrightarrow[-NH_3]{} \underset{\substack{\text{Sodium salt of}\\\text{carboxylic acid}}}{RCOO^{\ominus} \ \overset{\oplus}{Na}} \xrightarrow{H^{\oplus}} \underset{\text{Carboxylic acid}}{RCOOH}$$

Procedure: Take amide (1.0 g) and 10% aqueous sodium hydroxide solution (15.0 mL) in a clean 100 mL round bottomed flask fitted with water condenser. Add few pumice stones and heat the mixture over a wire gauze, gently for 30 minutes. Look for fumes of ammonia. Absence of ammonia smell, after heating for some time, indicates completion of reaction. Cool the solution in ice-water and add dropwise concentrated hydrochloric acid, until the mixture is acidic to litmus paper. Allow the mixture to stand in ice-water for a few minutes, and then filter the solid at the pump, wash with minimum amount of cold water. Dry the solid and recrystallize a part of crude sample using hot water or aqueous ethanol. Filter the crystallized solid, dry on a porous clay plate and record its melting point.

Note

- This method is useful only if acid derivative of amide is insoluble in water.

- Many amides and anilides require 70% sulphuric acid because of the higher solubility and higher permissible temperature in 70% sulphuric acid.

- In case the carboxylic acid formed on working up is water soluble, add 1-2 drops of phenolphthalein and neutralize the medium with 5% aqueous sodium hydroxide solution until pale pink solution is obtained. Use this solution to prepare the SBT derivative (see *section 4.3.2.1.1,* page 130).

Safe Waste Disposal

- Neutralize the filtrate with solid sodium carbonate (foam formation occurs, careful!) until pH ~7 and transfer to the aqueous solution container.

2. Oxalate or nitrate derivative for urea: Prepare a concentrated solution of urea (5.0 mL) and divide it into two portions.

(a) To the first part add concentrated nitric acid (1.0 mL) dropwise with shaking. White coloured crystalline precipitate of urea nitrate (melting point 163°C) seperates out. Filter the solid, wash with minimum amount of ice-cold water and dry. Take the melting point of the crystalline derivative (literature melting point 163°C).

(b) To the second part, dropwise add concentrated solution of oxalic acid (7-8 mL). White crystals of urea oxalate separate out (melting point, 171°C). Filter the solid, wash with minimum amount of ice-cold water and dry. Take the melting point of the crystalline derivative (literature metting point, 171°C).

$$NH_2CONH_2 \ + \ (COOH)_2 \ \longrightarrow \ NH_2CONH_2 \ (COOH)_2$$

Urea **Oxalic acid** **Urea oxalate**

Note

- Urea on heating with aqueous alkali followed by acidification liberates carbon dioxide and ammonia. Hence, urea nitrate and oxalates are prepared as derivatives.

$$NH_2CONH_2 \ + \ NaOH \longrightarrow NH_2COONa \ + \ NH_3$$
Urea

$$CO_2 \ + \ NH_3 \longleftarrow NH_2COOH$$

- Both urea nitrate and urea oxalate are salts of urea with nitric acid and oxalic acid respectively. Do not wash the crude precipitate with excess of water, else the solid will disappear.
- Urea nitrate is an explosive if left to dry, avoid its preparation in the lab.
- Imides are hydrolyzed to dicarboxylic acids with alkali followed by acidification. Use the same procedure described in method (1), page 231, to prepare derivative for imides.
- Nitriles can be converted into corresponding carboxylic acids by alkaline hydrolysis (Method 1) followed by acidification or by hydrolysis with concentrated sulphuric acid (See acid hydrolysis of anilides, page 236).

Post Lab Questions

1. Give one chemical test to distinguish between
   (a) Benzamide and urea
   (b) Benzamide and N-methylbenzamide
2. Benzamide does not react with 2,4-DNP but gives hydroxamic acid test. Explain.
3. How will you distinguish between the following pairs by IR spectroscopy:
   (a) Benzamide and N-methylbenzamide
   (b) Benzoic acid and benzamide

4. Give one application of biuret test in the lab other than in identification of urea.

5. Which will be more easily hydrolysed: Benzamide or methyl acetate? Give reason.

6. Alkaline hydrolysis of esters/amides is preferred over acidic hydrolysis. Give reasons.

7. A student is given an unknown compound which is a possible acid derivative (ester/amide/imide). How will he/she characterize the functional group based on the following:

   • Elemental analysis

   • Odour

   • Classification test with 10% aqueous NaOH

   • Reaction with alcoholic KOH.

4.3.2.6.3 Anilides

Anilides are secondary amides. Structurally, they are N-acyl derivatives of aniline, having general formulae RCONHAr or ArCONHAr. These compounds can be identified in the lab based on the products obtained on acid or alkaline hydrolysis. On hydrolysis, they yield aniline along with carboxylic acid or carboxylic acid salt depending upon the nature of catalyst used. Table 4.33 gives the classification tests for the anilides.

Table 4.33   Classification Tests for Anilides

S. No.	Classification Test	Test category	Observation	Inference
1.	Hydrolysis with acid	Group test	White precipitate of carboxylic acid (sodium bicarbonate test). Neutralization followed by +ve azo dye test.	Anilide confirmed
2.	Hydrolysis with alkali	Group test	Ether extract gives +ve azo dye test. Acidification of aqueous layer precipitates carboxylic acid (sodium bicarbonate test).	Anilide confirmed

Classification Tests for Anilides

1. Acid catalysed hydrolysis: Hydrolysis of anilides with 70% concentrated sulphuric gives carboxylic acid (may or may not be water soluble) and aniline which remains in the solution due to formation of anilinium salt. After removing the carboxylic acid precipitate (if insoluble), the medium is neutralised to precipitate aniline. Aniline can then be confirmed by the Azo dye test.

$$C_6H_5NH - \underset{\underset{O}{\|}}{C} - C_6H_5 \; + \; H_2O \; + \; HX \longrightarrow C_6H_5COOH \; + \; C_6H_5\overset{\oplus}{N}H_3 \; X^{\ominus}$$

Benzanilide   **Benzoic acid** **Anilinium salt**

X = Cl, Br, I

Procedure: Take the given compound (0.5 g) in a dry round bottomed flask and add 70% sulphuric acid (5.0 mL). Put 1-2 pumice stones and connect the condenser. Reflux the mixture for ~30-40 minutes. Cool the solution and filter the solid acid (if any). Wash the solid well with minimum amount of cold water and preserve the solid for sodium bicarbonate test. (See *section 4.3.2.1.1*, page 126, bicarbonate test). Neutralise the filtrate with dilute sodium hydroxide solution till neutral to pH paper. Remove the organic compound by extraction with ether and evaporate ether carefully. Perform the Azo dye test for aniline on the residue. (See *section 4.3.2.5.1*, for Azo dye test).

Note

- For preparation of 70% sulphuric acid, see Appendix II.
- In case of water-souble acid, neutralize the solution with dilute sodium hydroxide solution. Remove the aniline by extraction with ether (2×10 mL) and preserve the aqueous layer. Wash the ethereal layer with water, dry on anhydrous potassium carbonate and evaporate carefully. Perform the Azo dye test on the residue. Next, acidify the aqueous layer, concentrate to ~2.0 mL and prepare the SBT derivative. (See *section 4.3.2.1.1*, page 130)

Safe Waste Disposal

- Neutralize aqueous filtrate with 5% aqueous NaOH or 5% aqueous HCl as required until pH ~7. Transfer to aqueous solution container.

2. Alkaline hydrolysis: Alkaline hydrolysis of anilides will precipitate aniline and carboxylic acid salt remains in solution. Aniline is extracted with ether and ether evaporated carefully to collect the amine. The amine can then be subjected to Azo dye test (See *section 4.3.2.5.1*, page 209)

Neutralization of the filtrate with dilute sulphuric acid will precipitate the carboxylic acid, if it is insoluble in water. The precipitated acid may be identified by bicarbonate test. (See *section 4.3.2.1.1*, page 126, for bicarbonate test). Else concentrate the solution to ~2.0 mL and identify the acid from the solution by preparing an SBT derivative (see *section 4.3.2.1.1*, page 130, derivatives of carboxylic acids).

$$C_6H_5NH - \underset{\underset{O}{\|}}{C} - C_6H_5 \; + \; NaOH \longrightarrow C_6H_5COONa \; + \; C_6H_5NH_2$$

Benzanilide   **Sodium benzoate** **Aniline**

$$\downarrow H^{\oplus}$$

$$C_6H_5COOH$$

Benzoic acid

Procedure: Take the given compound (0.1 g) in a dry round bottomed flask and add 15% aqueous sodium hydroxide (15.0 mL). Add 1-2 pumice stones and reflux the mixture for ~30 minutes. Cool the solution and precipitate the amine (if solid). If liquid, extract with ether, evaporate the ether carefully and store the residue for Azo dye test. Else, wash the solid well with water and preserve the solid for Azo dye test. (See *section 4.3.2.5*). Neutralize the filtrate with dilute sulphuric acid solution till neutral to pH paper. Remove the organic compound by filtration (if water insoluble). Perform the bicarbonate test on the residue. (See *section 4.3.2.1.1,* page 126, for bicarbonate test.) Else prepare the SBT derivative after concentrating the solution to ~2.0 mL.

Note

- Anilides do not give hydroxamic acid test unlike most primary aromatic amides.

Safe Waste Disposal

- Neutralize the filtrate with 5% aqueous NaOH or 5% aqueous HCl as required until pH ~7. Transfer to the aqueous solution continer.

Derivatives of Anilides

1. Acid hydrolysis

Procedure: Take the given compound (1.0 g) and concentrated sulphuric acid (70%, 10.0 mL) in a clean and dry round bottomed flask fitted with a reflux condenser. Add 1-2 pumice stones. Connect the condenser and reflux on the wire gauze for ~30-40 minutes. Cool the contents then proceed as in (a) or (b).

(a) If solid separates on cooling, filter it. Wash well with minimum amount of cold water and recrystallize a fraction of crude product with hot water/aqueous ethanol. Record the melting point of the pure sample.

(b) If the acid is water soluble, no solid will precipitate. In such cases, extract the contents with ether (2 × 10 mL) and evaporate ether on rotatory evaporator. Prepare *S*-benzylisothiouronium carboxylate of the residue as per the procedure given in *section 4.3.2.1.1,* page 130.

Neutralize the filtrate from (a)/(b), after separation of acid, with dilute sodium hydroxide. Filter any solid so formed and determine its melting point after recrystallization with aqueous ethanol. Else, extract the amine (if liquid) with ether (2 × 10 mL) and evaporate the ether to dryness. Acylate or benzoylate the residual amine, recrystallize and record the melting point of the derivative amide. (For acetylation/benzoylation of amine, see *section 4.3.2.5,* page 213, derivatives of amines.)

Safe Waste Disposal

- Transfer any residual ethereal layer to the flammable organic waste container. Neutralize the aqueous filtrate with 5% aqueous HCl until neutral to pH. Then transfer to the aqueous solution container.

2. Bromo derivative: Anilides like acetanilide can be brominated to crystalline bromo derivatives. Bromination of some anilides can be carried out by using green method. (For detailed theory, see Volume 1, *section 6.3.4*).

O
‖
HN   CH$_3$

CAN, KBr → EtOH, H$_2$O

O
‖
HN   CH$_3$

Br

Acetanilide   ***p*-Bromoacetanilide**

Procedure: Dissolve a mixture of acetanilide (1.0 g), potassium bromide (1.0 g) in ethanol (20.0 mL) in a clean conical flask and mount the flask over a magnetic stirrer. Add dropwise, solution of ceric ammonium nitrate (6.0 g) in water (15.0 mL) to the conical flask, with stirring. Once the addition is complete, stir the reaction mixture for another 10 minutes on the magnetic stirrer. Thereafter, pour the contents into a beaker containing ice-cold water (30-40 mL). *p*-Bromoacetanilide will separate as a white solid. Filter at suction, dry and recrystallize a fraction of the crude sample with aqueous ethanol and record its melting point.

4.3.2.6.4   *Amino acids*

If your compound is aliphatic, neutral, cold-water soluble, contains nitrogen and gives effervescence with sodium bicarbonate, then it is most likely to be an amino acid. Most amino acids are amphoteric in nature and are characterized by the presence of both amino and carboxylic groups. The naturally occurring amino acids are mostly α-amino acids, containing amino group on the α-carbon to the carboxylic group. These compounds give reactions of both carboxylic group and primary aliphatic amines. Some reactions shown by amino acids are also due to combination of amino and carboxylic group. A list of classification tests for amino acids is given in Table 4.34.

Table 4.34   Classification Tests for Amino Acids

S. No.	Test	Category	Observation	Inference
1.	Ninhydrin test	Group test	Violet colouration[a]	Amino acid confirmed
2.	Complexation with copper sulphate	Special test	Blue-violet colouration	Amino acid confirmed
3.	Nitrous acid test[b]	Special test	Evolution of nitrogen gas	Primary aliphatic amine confirmed
4.	Bicarbonate test[b]	Special test	Brisk effervescence	Carboxylic group confirmed

a: Proline and hydroxy proline give yellow colouration; **b:** see individual tests for aliphatic amines and carboxylic acids, *sections 4.3.2.5, page 204* and *4.3.2.1.1, page 126*. respectively.

Classification tests for amino acids

1. Ninhydrin test: All α-amino acids react with Ninhydrin reagent, on heating to 110°C, to form violet coloured complex called **Ruhemann's purple**. Proline and hydroxy-proline give yellow coloration, being an α-imino acid.

Ninhydrin

Blue-violet colour complex
(Ruhemann Purple)

Procedure: In a clean test tube, take a solution of the given compound [0.02 g in water (1.0 mL)]. Add Ninhydrin reagent (1.0 mL) followed by addition of water (~10.0 mL). Heat the contents on a water bath for ~30 seconds. Development of blue-violet colour confirms an α-amino acid.

Note

- This test is used in the paper and thin layer chromatography to identify the amino acids through their R_f values.
- Proteins, peptides and some β-amino acids also give this test.

Safe Waste Disposal

- Transfer all the test solutions to the aqueous solution container.

2. Copper sulphate test: Amino acids combine with copper sulphate to give blue-violet coloured complex. This reaction is also basis of estimation of proteins and peptides in the laboratory (**Biuret test**). Water soluble amino acids give colouration in cold whereas water-insoluble amino acids give blue-violet colouration only on heating.

Amino acid

Copper Complex
(Blue)

Procedure: Take the solution of the given compound (0.02 g in 1.0 mL water) and add copper sulphate solution (3-4 drops). A blue coloured solution is formed in cold or heating in hot water bath for 4-5 minutes. Appearance of blue-violet colouration confirms the presence of amino acid.

Note

- Nitrous acid, carbylamine test and sodium bicarbonate test for amino acids may be performed as per the procedure given in *sections 4.3.2.5 and 4.3.2.1.1,* respectively.
- Use acetic acid in place of HCl while doing nitrous acid test for amino acids.
- α-Amino acids liberate nitrogen gas with formation of α-hydroxy acids, in the Nitrous acid test.

Safe Waste Disposal

- Place all the test solutions in the aqueous solution container.

3. Spot tests for amino acids

(a) *Ninhydrin test:* Spot 1-2 drops of the following solutions over a Whatman paper strip, sequentially, in the order detailed below:

- Concentrated solution of the compound
- Ninhydrin solution (2-3 times)
- Heat the strip in oven maintained at ~110°C for 2-3 minutes.

Development of blue-violet colour confirms the presence of amino acid.

Note

- Proline and hydroxy proline give yellow colouration.

4. Spectroscopic identification

Table 4.35 Spectral Data of Amino Acids

S.No.	Type of spectrum	Peak value	Peak attribution	Remarks
1.	IR	1400 & 1600 cm^{-1}	>C=O Str	Doublet
		2600-3300 cm^{-1}	N-H str(NH_3^+)	
		1400-1660	N-H bending	
2.	NMR	7.5 δ	NH_3^+	Broad peak
		4.2-4.5 δ	-C$\underline{H}$R-NH_3^+	

Derivatives of Amino Acids

1. Benzoyl derivative: The benzoyl derivative of a given amino acid can be obtained *via* **Schotten Baumann reaction,** using benzoyl chloride in the presence of aqueous sodium hydroxide. Due to the presence of carboxylic group in amino acids, the benzoyl derivative forms the corresponding sodium salt and remains soluble in water and can be separated from the reaction mixture only after acidification. (for detailed theory on benzoylation, see Volume 1, *section 6.8.3*)

$$NH_2CHRCOOH + C_6H_5COCl \xrightarrow{NaOH} C_6H_5CONHCHRCOONa$$

Amino acid   Benzoyl chloride

$$C_6H_5CONHCHRCOONa \xrightarrow{conc.\ HCl} C_6H_5CONHCHRCOOH + NaCl$$

Benzoyl derivative

Procedure: Place the given amino acid (1.0 g) and 10% aqueous NaOH (20.0 mL) in a conical flask fitted with a stopper. Add first lot of benzoyl chloride (1.5 mL, added in two lots of 0.75 mL each) and stopper the flask immediately. Shake the reaction mixture vigorously for 10-15 minutes or till contents come back to room temperature. Add the second lot of benzoyl chloride and repeat as stated above. Allow the contents to come back to room temperature. Cool the flask by adding ice-water (~50.0 mL) and acidify the reaction mixture with concentrated HCl. Filter the separated solid at suction pump, wash well with minimum amount of cold water. Recrystallize a part of the crude sample with aqueous alcohol. Filter the crystallized solid, dry on a porous clay plate and record its melting point.

Note

- Completion of reaction is indicated by attainment of room temperature by the contents and also by disappearance of benzoyl chloride smell.

Safe Waste Disposal

- Transfer the filtrate after neutralization with 5% aqueous NaOH to aqueous solution container.

2. 3,5-Dinitrobenzoyl derivative: On reaction with 3,5-dinitrobenzoyl chloride followed by acidification with concentrated acid, amino acids form 3,5-dinitrobenzoyl derivative.

3,5-Dinitrobenzoyl derivative

Procedure: Place the given amino acid (1.0 g) and 10% aqueous NaOH (20.0 mL) in a conical flask fitted with a stopper. Add first lot of 3,5-dinitrobenzoyl chloride (1.5 mL added in two lots of 0.75 mL each; See Appendix II for preparation) and stopper the flask immediately. Shake the reaction mixture vigorously for 10-15 minutes or till contents come back to room temperature. Add the second lot of 3,5-dinitrobenzoyl chloride and repeat as stated above. Allow the contents to come back to room temperature. Filter the undissolved solid and acidify the reaction mixture with dilute aqueous HCl. Filter the separated solid on suction pump. Recyrstallize a part of the crude sample using aqueous alcohol or hot water. Filter the crystallized solid, dry on a porous clay plate and record its melting point.

Safe Waste Disposal

- Neutralize the aqueous filtrate with solid sodium carbonate (foam) until pH ~7. Transfer to the aqueous solution container.

Post Lab Questions

1. Amino acids are soluble in both acids and bases. Explain.
2. Most naturally occurring amino acids give purple coloration with Ninhydrin except proline and hydroxyl proline. Explain.
3. Aniline and phenylalanine are subjected to benzoylation reaction. What special precaution will you take in latter case and why?
4. Suggest a method for spectrophotometric estimation of amino acids in solution.
5. How will you differentiate between aniline and phenyl alanine by :
   (a) Chemical method(any two)
   (b) Physical Method

APPENDIX I

Classified Table of Organic Compounds

Table I   Aliphatic Carboxylic Acids and Derivatives

Name of the acid	M.Pt. (°C)	B.Pt. (°C)	S-Benzyl isothiouronium salt M.Pt. (°C)	Amide M.Pt. (°C)	Anilide M.Pt. (°C)	Anhydride M.Pt. (°C)
Formic acid	8	100	152	3	50	
Acetic acid	17	118	134	82	113	
Glycolic acid	79		146	120	97	
Propionic acid		140	153	79	105	
n-Butyric acid		163	146	115	96	
Iso-Butyric acid		155	143	129	105	
Pyruvic acid	13	165	158	124	104	
Monochloroacetic acid	63	189	160	119	137	46
Dichloroacetic acid		194	178	97	119	
Trichloroacetic acid	58	197	149	141	94	
Bromoacetic acid	50	208	145	91	130	42
Cyano acetic acid	66		-	123	198	-
Carbonic acid		90		132	238	
Oxalic acid	101*		196	418^d	245	
Malonic acid		133^d	146^d	170	225	
Succinic acid	185	235^d	154	242	226	120
Glutaric acid	98	303	153	174	224	56
Adipic acid	153	115	163	220	238	22
dl-Lactic acid	18	119	153	74	59	

(Contd.)

l-Malic acid	100		158	157	197	
d-Tartaric acid	169		-	196[d]	264[d]	
Citric acid	75-80**				210[d]	199	
Maleic acid	132		173	181[d]	187	56

*dihydrate; anhydrous, M.Pt. 190°C

**Monohydrate; anhydrous, M.Pt. 153°.

Table II   Aromatic Carboxylic Acid and Derivatives

Name of the Acid	M. Pt. (°C)	S-Benzyl isothiouronium salt M.Pt. (°C)	Amide M.Pt. (°C)	Anilide M.Pt. (°C)
Benzoic acid	121	167	129	162
Salicylic acid	158	148	139	135
Acetylsalicylic acid	135	144	138	136
m-Hydroxybenzoic acid	201	162	167	157
p-Hydroxybenzoic acid	213	145	162	197
o-Chlorobenzoic acid	141	169	141	118
m-Chlorobenzoic acid	158	164	134	124
p-Chlorobenzoic acid	243	194	179	194
o-Bromobenzoic acid	150	171	155	141
m-Bromobenzoic acid	155	168	155	146
o-Methoxybenzoic acid	101	164	129	131
m-Methoxybenzoic acid	110	176	134	-
p- Methoxybenzoic acid	184	185	163	171
o-Nitrobenzoic acid	147	159	175	161
m- Nitrobenzoic acid	141	163	142	154
p- Nitrobenzoic acid	239	182	200	211
2,4-Dinitrobenzoic acid	183	181	204	193
3,5-Dinitrobenzoic acid	207	179	183	234
Anthranilic acid	146	144	109	131
m-Aminobenzoic acid	174	160	111[1]	140
p- Aminobenzoic acid	188	166	179[2]	162
o-Toluic acid	105	145	143	125
m-Toluic acid	111	164	95	126
p- Toluic acid	178	190	159	148
Phthalic acid	196[d,3]	158	220	251
Terephthalic acid	300[sub.]	206	>250	337
Phenylacetic acid	76	165	157	118

sub. = sublime; d = decompose   *(Contd.)*

Cinnamic acid	133	183	147	153
o-Nitrocinnamic acid	240	-	185	-
dl-Mandelic acid	118	166	134	152
1-Naphthoic acid	162	149	202	163
2-Naphthoic acid	185	184	192	170
Hexahydrobenzoic acid	30	166	185	146

[1]Hydrate, M.Pt. 79°. [2]Hydrate, M.Pt. 114°C.

[3]Phthalic anhydride M.Pt. 132°C

Table III   Phenols and Derivatives

Name of the Phenol	M.Pt. (°C)	B.Pt. (°C)	Benzoate M.Pt. (°C)	Bromo dvt. M.Pt. (°C)	3,5-Dinitro-benzoates M.Pt. (°C)
Phenol	43	182	69	95 (tri)	146
o-Cresol	30	191	-	56 (di)	138
m-Cresol	12	202	55	84(tri)	165
p-Cresol	36	202	70	199 (tetra) 49 (di)	189
o-Chlorophenol		176			143
m-Chlorophenol	31	214	71	-	156
p-Chlorophenol	43	220	88	-	186
o-Bromophenol	5	195	-	95	-
m-Bromophenol	33	236	86	-	-
p-Bromophenol	64	238	102	95	191
o-Iodophenol	43				
m-Iodophenol	40		73		183
p-Iodophenol	94		119	-	
Catechol	105	245	84 (di)	192 (tetra)	152
Resorcinol	110	280	117 (di)	112 (di)	201(di)
Hydroquinone	170	286		186 (di)	317(di)
o-Methoxyphenol	32	205	57	116 (tri)	142
m-Methoxyphenol		244	-	104 (tri)	-
p- Methoxyphenol	56	243	87		166
o-Nitrophenol	45	216	59	117 (di)	155
m-Nitrophenol	97	194/70	95	91 (di)	159
p-Nitrophenol	114		142	142 (di)	188
Salicylic acid	159				

(Contd.)

2,4-Dichlorophenol	45	210	97	68	-
2,4-Dibromophenol	40	239	97	95	-
2,4,6-Trichlorophenol	69	246	75	-	136
2,4,6-Tribromophenol	94		81	120 (tetra)	174
Picric Acid	122		-	-	-
Pyrogallol (1, 2, 3)	133	309	90 (tri)	158	205 (tri)
Phlorglucinol (1,3,5)	218		174 (tri)	151 (tri)	162 (tri)
1-Naphthol	94	279	56	105 (di)	217
2-Naphthol	123	285	107	84	210

Table IV   Aliphatic Aldehydes and Derivatives

Name of the Aldehyde	B.Pt. (°C)	2, 4- Dinitrophenyl-hydrazone M.Pt. (°C)	Semicarbazone M.Pt. (°C)	Oxime M.Pt. (°C)
Formaldehyde		166°	169d.°	-
Propionaldehyde	49	155	89,154[1]	40
n-Butyraldehyde	74	126	95,106[1]	-
Iso- Butyraldehyde	64	187	125	-
Aldol	77/16		194	
Crotonaldehyde	104	190	201	119
n-Valeraldehyde	103	108	-	52
n-Hexaldehyde	131	107	109	51
n-Heptaldehyde	156	108	109	57
n-Octaldehyde	170	106	101	
Chloral	98	131	90d	56
Bromal	174		87d	
Furfural	161	229	202	89
Tetrahydrofurfural	145	204	166	-
Hexahydrobenzaldehyde	162	172	173	

Table V   Aromatic Aldehydes and Derivatives

Name of the Aldehyde	M.Pt. (°C)	B.Pt. (°C)	2, 4-Dinitrophenyl-hydrazone M.Pt. (°C)	Semicarbazone M.Pt. (°C)	Oxime M.Pt. (°C)
Benzaldehyde		179	237	222	35
o-Chlorobenzaldehyde	11	213	207	225	75
m-Chlorobenzaldehyde	17	213	248	228	70

(Contd.)

p-Chlorobenzaldehyde	47	214	270	230	110
o-Bromobenzaldehyde	22	230	-	214	
m-Bromobenzaldehyde		234	25	205	
p-Bromobenzaldehyde	57		260	228	129
o-Nitrobenzaldehyde	44	252	250^d	256	103
m-Nitrobenzaldehyde	58		293^d	246	118
p-Nitrobenzaldehyde	106		320	221	129
o-Aminobenzaldehyde	40		250	247	
m-Aminobenzaldehyde			270^d	$280^{d.}$	
p-Aminobenzaldehyde	72		194	$173^{d.}$	
Salicylaldehyde		196	252^d	231	57
m-Hydroxybenzaldehyde	104	240	260^d	198	88
p-Hydroxybenzaldehyde	115		280^d	224	73
o-Methoxybenzaldehyde		243	253	215	
m-Methoxybenzaldehyde	4	230	219	233^d	
p-Methoxybenzaldehyde /Anisaldehyde	2.5	248	254^d	210	-
o-Tolualdehyde		200	193	212	49
m-Tolualdehyde		199	212	216	60
p-Tolualdehyde		204	234	234	80
4-Hydroxy-3-methoxy-benazldehyde/Vanillin	80	285^d	271	229^d	117
Veratraldehyde/3,4-Dimethoxybenzaldehyde	44	285	265	177	
p-Dimethylamino-benzaldehyde	74	176	237	222	
Phenylacetaldehyde	34	194	121	156	103
Cinnamaldehyde		252^d	255^d	215	138
Dihydrocinnamaldehyde		224	149	127	138
1-Naphthaldehyde	34	292	270	221	90
2- Naphthaldehyde	60			245	156

Table VI Aliphatic and Alicyclic Ketones and Derivatives

Name of the Ketone	B.Pt. (°C)	2, 4- Dinitrophenyl-hydrazone M.Pt. (°C)	Semicarbazone M.Pt. (°C)	Oxime M.Pt. (°C)
Acetone	56	126	190	59
Ethyl methyl ketone	80	115	146	-

(Contd.)

Diethyl ketone	102	156	139	69
Methyl propyl ketone	102	144	112	58
Methyl isopropyl ketone	94	120	113	109
Di- *n*-propyl ketone	144	75	133	-
Di-iso-propyl ketone	124	98	160	34
n-Butyl methyl ketone	128	106	122	49
Isobutyl methyl ketone	117	95	132	58
Di-*n*-butyl ketone	188	-	90	-
Di-iso-butyl ketone	168	66	122	-
Diacetyl	88	315 (di)	279 (di)	234 (di)
Acetylacetone	139	209 (di)	-	149 (di)
Acetonylacetone	194	257	224	138 (di)
Mesityl oxide	130	203	164	49
Ethyl acetoacetate	181	96	133	-
Cyclopentanone	131	147	209	56
Cyclohexanone	156	162	166	90
2-Methyl cyclohexanone	165	137	197	43
3-Methyl cyclohexanone	170	155	190	-
4-Methyl cyclohexanone	171	134	199	38
Cycloheptanone	180	148	163	-

Table VII   Aromatic Ketones and Derivatives

Name of the Ketone	M.Pt. (°C)	B.Pt. (°C)	2, 4-Dinitrophenyl-hydrazone M.Pt. (°C)	Semicarbazone M.Pt. (°C)	Oxime M.Pt. (°C)
Acetophenone	20	202	250	198	59
o-Chloroacetophenone		229	206	160	-
m-Chloroacetophenone		228		232	-
p-Chloroacetophenone	20	232	231	200	95
o-Bromoacetophenone		112/10	189	177	-
m-Bromoacetophenone		131/16		238	-
p-Bromoacetophenone	51	256	230	208	128
ω-Bromoacetophenone	50		-	146	97
o-Hydroxyacetophenone	28	215	-	210	117
m-Hydroxyacetophenone	96	296	-	195	-
p-Hydroxyacetophenone	109	240	261	199	145

(Contd.)

o-Methoxyacetophenone		245	159	183	-
m-Methoxyacetophenone		240	207	196	-
p-Methoxyacetophenone	38	258	220	198	87
o-Nitroacetophenone	28	218	-	-	-
m-Nitroacetophenone	81	202	228	257	132
p-Nitroacetophenone	81		-	-	174
Propiophenone	19	218	191	181	-
Benzyl methyl ketone	27	216	156	198	69
Benzyl ethyl ketone		226	-	135	-
Benzophenone	49	306	239	164	142
p-Chlorobenzophenone	78	323	185	-	163
p-Bromobenzophenone	82	350	230	-	-
Phenyl *p*-tolyl ketone	60	326	199	121	-
Benzyl phenyl ketone	60	320	204	148	-
Di-*p*-tolyl ketone	95	335	229	140	-
Dibenzyl ketone	35	331	100	146	124
Methyl-1-naphthyl ketone	34	302	-	229	137
Methyl-2-naphthyl ketone	54	301	262	236	-
Benzoin	137	344	245	206d.	151
Benzil	95	347d.	- 189	182 (mono) 244 (di)	137 (mono) 237 (di)
Phenacyl chloride	59	244	212	156	-
Phenacyl bromide	50	135/18	230	146	-
p-Bromophenacyl bromide	109		-	-	115
Benzalacetone	42	262	227	187	115
Dibenzalacetone	112		180	190	143
1-Tetralone		129/12	257	217	-
Fluorenone	83	341	283	234	196

Table VIII Esters

Name of Ester	B.Pt. (°C)	Acid derivative M.Pt. (°C)
Ethyl benzoate	212	122
Methyl-*o*-toluate	213	104
Methyl-*m*-toluate	215	110
Diethyl oxalate	186	101

(Contd.)

Di-isopropyl oxalate	190	101
Methyl salicylate	224	159
Ethyl salicylate	234	159
Isopropyl salicylate	237	159
n-Propyl salicylate	238	159
Diethyl maleate	224	130
n-Propyl benzoate	230	122
Phenyl benzoate	68	118
1-Napthyl benzoate	56	122
2-Napthyl benzoate	107	122
Resorcinol dibenzoate	117	122
1- Naphthyl acetate	49	118
2-Naphthyl acetate	71	118
Phenyl Salicylate	42	159
Benzyl Cinnamate	39	133
Diphenyl succinate	121	185

Table IX   Alcohols and Derivatives

Name of the alcohol	M.Pt. ($^{\circ}C$)	B.Pt. ($^{\circ}C$)	3,5-Dinitrobenzoate M.Pt. ($^{\circ}C$)	p-Nitrobenzoate M.Pt. ($^{\circ}C$)
Methyl alcohol		64.5	109	960
Ethyl alcohol		78	94	57
n-Propyl alcohol		97	75	35
Iso-Propyl alcohol		82	122	110
Allyl alcohol		97	50	30
n-Butyl alcohol		118	64	35
Iso-Butyl alcohol		108	88	69
Sec-Butyl alcohol		100	76	26
Tert-Butyl alcohol	25	82	142	116
n-Amyl alcohol		138	46	11
2-Pentanol		119	62	17
3-Pentanol		116	100	17
n-Hexanol		156	61	5
Hexan-2-ol		140	39	40
n-Heptanol		176	48	10
Heptan-2-ol		159	49	
n-Octanol		194	62	12

(Contd.)

Octan-2-ol		179	32	28
n-Nonyl alcohol		214	52	66
n-Decyl alcohol	6	231	57	30
Cyclopentanol		141	115	62
Cylclohexanol	25	161	113	50
Ethylene glycol		197	169 (di)	140 (di)
Propylene glycol		187	147 (di)	127 (di)
Aryl-Substituted Alcohols and Derivatives				
Benzyl alcohol		205	113	85
o-Chlorobenzyl alcohol	74			94
p-Bromobenzyl alcohol	77			121
p-Nitrobenzyl alcohol	93	185/12		171
o-Methoxybenzyl alcohol		249		82
m-Methoxybenzyl alcohol		252	121	
p-Methoxybenzyl alcohol	25	259		94
1-Phenylethyl alcohol	20	203	95	48
2-Phenylethyl alcohol		220	108	62
Diphenylcarbinol	69	299	141	131
Cinnamyl alcohol	33	257	121	78
Benzoin	137	344		123

Table X   Carbohydrate and Derivatives

Name of the carbohydrate	M.Pt. (°C)	Osazone M.Pt. (°C)	Acetate M.Pt. (°C)
D-Ribose	87	163	-
D-Fructose	102	205	70 (α-form) 109 (β-form)
Maltose	102 (monohydrate) 160-165 (anhydrous)	208	125 (α-form) 159 (β-form)
Raffinose	118 (anhydrous) 78 (pentahydrate)	-	99 (β-form)
D-Mannose	132	205	74 (α-form) 115 (β-form)
α-*D*-Xylose	145	163	59 (α-form) 126 (β-form)
α-*D*-Glucose	85 (monohydrate) 146 (anhydrous)	205	112 (α-form) 134 (β-form)

(Contd.)

β-*D*-Arabinose	160	166	96 (α-form) 86 (β-form)
D-Galactose	119 (monohydrate) 168 (anhydrous)	196	96 (α-form) 142 (β-form)
Sucrose	169	205	82 (β-form)
Lactose	203 (monohydrate) 223 (anhydrous)	200	152 (α-form) 100 (β-form)

Table XI   HaloHydrocarbons and Derivatives

Name of Halohydrocarbon	M.Pt. (°C)	B.Pt. (°C)	Nitro dvt. M.Pt. (°C)
Benzyl chloride		179	1-(Chloromethyl)-4-nitrobenzene (71°C)
Chlorobenzene		132	2,4-Dinitrochlorobenzene (51°C)
Bromobenzene		156	2,4-Dinitrobromobenzene (75°C)
Iodobenzene		188	4-Iodo-nitrobenzene (171°C)
o-Dichlorobenzene		179	1,2-Dichloro-4,5-dinitrobenzene (110°C)
p-Dichlorobenzene	53	-	1,4-Dichloro-2-nitrobenzene (54°C)
o-Dibromobenzene		224	1,2-Dibromo-4,5-dinitrobenzene (114°C)
m-Dibromobenzene		219	2,4-Dibromo-1-nitrobenzene (114°C)
p-Dibromobenzene	89		1,4-Dibromo-2-nitrobenzene (84°C)
o-Chlorotoluene		159	1,2-Dichloro-3,5-dinitrobenzene (63°C)
m-Chlorotoluene		162	1-Chloro-5-methyl-2,4-dinitrobenzene (103°C)
p-Chlorotoluene		162	1-Chloro-4-methyl-2-nitrobenzene (38°C)
m-Bromotoluene		183	1-Bromo-5-methyl-2,4-dinitrobenzene (103°C)
p-Bromotoluene	28	185	1-Bromo-4-methyl-2-nitrobenzene (47 °C)
o-Iodotoluene		207	1-Iodo-2-methyl-4-nitrobenzene (103°C)

Table XII   Aromatic Hydrocarbons and Derivatives

Name of Aromatic hydrocarbon	M.Pt. (°C)	B.Pt. (°C)	Nitro dvt. (M.Pt.)	Picrate (°C)
Benzene	5	81	1,3-Dinitrobenzene (89°C)	84
Toluene	−95	111	2,4-Dinitrotoluene (70°C)	88
o-Xylene	−25	144	1,2-Dimethyl-4,5-dinitrobenzene (118°C)	81
m-Xylene	−47	139	2,4-Dimethyl-1,3,5-trinitrobenzene (183°C)	91
p-Xylene	15	138	1,4-Dimethyl-2,3,5-trinitrobenzene (139°C)	90
Mesitylene	−57	165	1,3,5-Trimethyl-2,4,6-trinitrobenzene (235°C)	97
Ethylbenzene	−94	136	2,4,6-Trinitroethylbenzene (37 °C)	96

(Contd.)

Stilbene	125	306	-	95
Naphthalene	80	218	1-Nitronaphthalene (61°C)	152
Anthracene	216	351	-	135
Phenanthrene	98	340	-	144
Biphenyl	69	254	4,4'-Dinitro-1,1'-biphenyl (237 °C)	-
Diphenylmethane	25	261	Bis(2,4-dinitrophenyl)methane (172°C)	-
Triphenylmethane	92	358	Tris(4-nitrophenyl)methane (206°C)	-

Table XIII Primary Amines and Derivatives

Name of Amine	M.Pt. (°C)	B.Pt. (°C)	Picrate M.Pt. (°C)	Acetyl dvt. M.Pt. (°C)	Benzoyl dvt. M.Pt. (°C)	Any other
Allylamine		56	140	-	17	
n-Butylamine		77	151	-	42	
Iso-Butylamine		69	156	-	57	
Sec-Butylamine		63	140	-	76	
n-Amylamine		105	143	-	-	
n-Hexylamine		129	127	-	40	
n-Heptylamine		155	121	-		
Benzylamine		185	199	60	105	
1-Phenylethylamine		187	189	57	120	
2-Phenylethylamine		198	167	-	116	
Cyclohexylamine		134	158	104	149	
Aniline		184	-	114	163	Tribromo dvt. (119°C)
o-Toluidine		200	214	112	143	Dibromo dvt. (50°C)
m-Toluidine		203	200	65	125	
p-Toluidine	45	200	181	153	158	Dibromo dvt. (73°C)
o-Chloroaniline		209	134	87	99	
m-Chloroaniline		230	177	72	122	
p-Chloroaniline	71	232	177	177	192	
o-Bromoaniline	32	229	129	99	116	
p-Bromoaniline	66	245	180	167	204	
o-Iodoaniline	60		180	167	139	
m-Iodoaniline	30		112	119	157	

(Contd.)

p-Iodoaniline	64		-	184	222	
o-Nitroaniline	72		73	94	98	
m-Nitroaniline	114	225	143	155	157	
p-Nitroaniline	148		100	216	199	
o-Anisidine	5	225	200	85	60	
p-Anisidine	57	246	117	127	153	
o-Phenetidine	-	229	119	79	104	
m-Phenetidine	-	248	158	96	103	
p-Phenetidine	-	254	69	138	173	
o-Aminophenol	174	-	-	124(di)	184 (di)	
m-Aminophenol	123	-	-	101(di)	153 (di)	
p-Aminophenol	186	-	-	151(di)	234 (di)	
2,4-Dichloroaniline	63	245	106	146	117	
2,4-Dibromoaniline	79		124	146	134	
2,4,6-Trichloroaniline	78	262	83	206	174	
2,4,6-Tribromoaniline	120	300	-	238	198	
1-Naphthylamine	50	300	163	160	161	
2-Naphthylamine	113	294	195	134	162	
o-Phenylenediamine	102	257	208	186	303(di)	
m- Phenylenediamine	64	283	184	191	240(di)	Dibromo dvt. (158°C)
p- Phenylenediamine	141	267	202	304	300(di)	
p-Amino-diethylaniline		261	-	104	172	

Table XIV   Secondary Amines and Derivatives

Name of Amine	M.Pt. (°C)	B.Pt. (°C)	Picrate M.Pt. (°C)	Acetyl dvt. M.Pt. (°C)	Benzoyl dvt. M.Pt. (°C)
Diethylamine		56	155	-	42
Di-*n*-Propylamine		111	75	-	-
Di-iso-Propylamine		84	140	-	-
Di-*n*-Butylamine		159	59	-	
Di-iso-Butylamine		139	119	86	
N-Methylaniline		194	145	103	63
N-Ethylaniline		206	138	54	60
N-*n*-propyl-aniline		222		47	
Diphenylamine	54	302	182	103	180

Table XV Teritiary Amines and Derivatives

Name of Amine	M.Pt. (°C)	B.Pt. (°C)	Picrate M.Pt. (°C)
Triethylamine		90	173
Tri-*n*-propylamine		156	117
Tri-*n*-butylamine		155	106
Tricyclohexylamine		193	163
N,N-Ethylmethylaniline		201	134
N,N-Diethylaniline		216	142
Di-*n*-Propylaniline		245	261
N,N-Dibenzylaniline	70	300	132
N,N-Dimethyl-*o*-toluidine		185	122
N,N-Dimethyl-*m*-toluidine		212	131
N,N-Dimethyl-*p*-toluidine		211	130
N,N-Dimethyl-2-naphtylamine	47	305	206
Pyridine		115	167
5-Methylquinoline	49	263	218

Table XVI Nitro–Compounds and Derivatives

Name of Nitro Compound	M.Pt. (°C)	B.Pt. (°C)	Nitro dvt./Reduction product M.Pt
m-Dinitrobenzene	90	302	*m*-Nitroaniline (114°C)/ *m*-phenylenediamine (63°C)
Nitrobenzene	5	216	*m*-Dinitrobenzene (90°C)
o-Nitrotoluene	Liq.	220	2,4-Dinitrotoluene (70°C)
p-Nitrotoluene	52	238	2,4-Dinitrotoluene (70°C)
o-Nitrophenol	46	216	See phenols
m-Nitrophenol	97		See phenols
p-Nitrophenol	114		See phenols
o-Nitrobenzaldehyde	44		See Aldehydes
m-Nitrobenzaldehyde	58		See Aldehydes
p-Nitrobenzaldehyde	106		See Aldehydes
o-Nitrobenzyl chloride	48		2,4-Dinitrochlorobenzyl chloride (34°C)
m-Nitrobenzyl alcohol	27		See Alcohol
o-Chloronitrobenzene	33	244	2,4-Dinitro chlorobenzene (51°C)
m-Chloronitrobenzene	45	236	3,4-Dinitro chlorobenzene (36°C)
p-Bromonitrobenzene	126	259	2,4-Dinitrobromobenzene (74°C)/ *p*-Bromoaniline (66°C)

(Contd.)

2,4-Dinitrochlorobenzene	51	315	2,4,6-Trinitrochlorobenzene/ Picryl chloride (83°C)
2,4–Dinitrobromobenzene	75		*p*-Phenylenediamine (140°C)/ *p*-Nitroaniline (147°C)
o-Nitro-benzoic acid	148		See Carboxylic acids
m-Nitrobenzoic acid	141		See Carboxylic acids
p-Nitrobenzoic acid	241		See Carboxylic acids
1-Nitronaphthalene	61	304	1,3,8-Trinitronaphthalene (218°C)
1,5-Nitronaphthalene	214		1,4,5-Trinitronaphthalene (154°C) 1,5-diaminonaphthalene (190°C)

Table XVII   Amides, Anilides, Urea and Derivatives

(for Amides, see under carboxylic acids)

Name of Amides	M.pt. (°C)	Derivative (M.Pt.)
Acetamide	82	Picrate (107 °C) Anilide (114 °C)
m-Toluamide	95	Anilide (126 °C) *m*-Toluic acid (110 °C)
Acetanilide	114	*p*-Nitroacetanilide (210 °C) *p*-Bromoacetanilide (167 °C)
Benzamide	128	Benzoic acid (122°C)
m-Nitroacetanilide	154	*m*-Nitroaniline (114°C)
Benzanilide	164	*p*-Bromobenzanilide (204°C)
p-Nitroacetanilide	210	*p*-Phenylenediamine (140°C)
p-Nitrobenzamide	201	*p*-Nitrobenzoic acid (240°C)
Urea	132	Urea oxalate (171°C), Urea nitrate (163°C)

Table XVIII   Amino Acids and Derivatives

Name of Amino acid	M.Pt. (°C)	Benzoyl dvt. M.Pt. (°C)	3,5-Dintrobenzoyl Dvt. M.Pt. (°C)
Glycine	232^d	187	179
Alanine (dl)	295^d	165	177
Phenylglycine	127	63	-
Hippuric acid	187	Amide (183°C) Anilide (208°C)	
Anthranilic acid	144	181	278
m-Aminobenzoic	174	248	270
p-Aminobenzoic acid	186	278	>290

Reagent Preparations

1. **Acetate buffer (pH-5.6):**

   (a) Prepare acetic acid (0.2M): In a clean reagent bottle (500 mL), dissolve glacial acetic acid (6.0 mL) in distilled water (500 mL). Shake and stopper the reagent bottle.

   (b) Prepare sodium acetate (0.2 M): In a clean reagent bottle (500 mL), dissolve anhydrous sodium acetate (8.2 g) in distilled water (500 mL). Shake and stopper the reagent bottle.

   (c) In a clean standard volumetric flask (1000 mL), add acetic acid (4.8 mL, 0.2 M) and sodium acetate (452 mL, 0.2 M). Make up the volume up to the mark with distilled water. Shake and stopper the reagent bottle.

2. **2% Alcoholic silver nitrate:** In a clean amber coloured bottle, dissolve silver nitrate (2.0 g) in ethanol (100 mL). Shake and tightly stopper the reagent bottle for future use.

3. **Alcoholic potassium hydroxide (0.5N):** Dissolve anhydrous potassium hydroxide (AR, 10.0 g) in 95% ethanol (250.0 mL) with stirring. Allow the solution to settle overnight. Filter any suspended particles from alcoholic potassium hydroxide solution kept overnight, before use.

4. **Alkali iodide-azide:** In a clean and dry conical flask, dissolve NaOH (40.0 g) and KI (20.0 g) in distilled water (70.0 mL). Stir on a magnetic stirrer, warm if required. Then add a solution of NaN_3 (1.0 g) in distilled water (10.0 mL) and dilute to 100 mL with distilled water.

5. **α-Amylase solution:** Dissolve α-amylase (100.0 mg) in acetate buffer (100.0 mL, pH 5.6) to make stock solution. Store in refrigerator. Dilute 1:50 with acetate buffer to obtain working solution whenever required.

6. **Analytical reagent for Lowry's method:** Take 1% aqueous $CuSO_4$ (0.5 mL) in a clean and dry standard flask (50.0 mL). Add 2% aqueous Rochelle salt (0.5 mL) followed by addition of 2% Na_2CO_3 solution in 0.1M NaOH (49.0 mL). Homogenize the solution, use same day.

7. **Aqueous ammonia (5% ammonium hydroxide solution):** Dilute concentrated ammonia (1 volume) with water (3 Volume) and store in tightly corked reagent bottle.

8. **Baeyer's reagent (2%):** In a clean reagent bottle (100 mL), dissolve potassium permanganate (2.0 g) in distilled water (100 mL) and add anhydrous sodium carbonate (10.0 g). Shake to dissolve and store.

9. **Barfoed reagent:** In a clean reagent bottle (200 mL), dissolve copper acetate (13.3 g) in 1% acetic acid (200 mL). Filter, if required and store.

10. **Benedict reagent:** Dissolve anhydrous sodium carbonate (10.0 g) and sodium citrate (17.3 g) in water (70.0 mL) taken in a clean reagent bottle (100 mL). Add a solution of hydrated copper sulphate (1.75 g) in water (10.0 mL) to the same reagent bottle and make up the volume to 100 mL.

11. **Brady's reagent (Aqueous alcoholic solution of 2,4-DNP):** In a clean conical flask (100 mL), dissolve 2,4–dinitrophenyl hydrazine (1.0 g) in concentrated sulphuric acid (5.0 mL). Add to this, slowly with shaking, methanol (25.0 mL). Cool and dilute carefully with water (10.0 mL). Allow the solution to stand for 5-6 hours. Filter to remove any precipitate and store in amber coloured bottle. This reagent can be used for preparation of 2,4-dinitrophenylhydrazones of both water insoluble and water soluble carbonyl compounds.

12. **Bromine in acetic acid (10%, v/v):** Dissolve bromine (10.0 mL) in acetic acid (100 mL) in a clean reagent bottle. Shake well, stopper and store.

13. **Bromine in carbon tetrachloride (10%, v/v):** Dissolve bromine (1.0 mL) in carbon tetrachloride (10.0 mL). Swirl well and use the brown liquid. This reagent may be prepared fresh as and when required.

14. **Bromine in water (5%, v/v):** Dissolve bromine (5.0 mL) in distilled water (100 mL) in a clean reagent bottle. Swirl well, stopper and store.

15. **BSA solution:** Dissolve BSA (0.02 g, Biochemistry grade) in distilled water (100 mL) and store. This reagent is stable for 1 month (avoid frothing!).

16. **Calcium chloride solution:** In a clean reagent bottle (1000 mL), dissolve anhydrous calcium chloride (50.0 g or hydrated calcium chloride; 100.0 g) in distilled water (1000 mL). Shake well and store.

17. **Ceric ammonium nitrate reagent:** In a clean conical flask, warm nitric acid (2N, 100 mL) and dissolve ceric ammonium nitrate (40.0 g) in it. Transfer to a clean reagent bottle (100 mL) and store. This reagent is stable for one month.

18. **3,5-Dinitrobenzoyl chloride:** Take 3,5-dinitrobenzoic acid (1.0 g) in a clean and dry china dish. Add phosphorous pentachloride (4.0 g) and grind together, until the mixture liquifies. If the mixture does not liquefy after grinding for 10-15 minutes, warm on a steam water bath (use Fume hood!). Use the liquid 3,5-dinitrobenzoyl chloride formed, immediately.

19. **3,5-Dinitrosalicylic acid (DNSA) reagent:** In a clean conical flask (100 mL), dissolve DNSA (1.0 g) in 2M NaOH (20.0 mL). Mount the conical flask on a magnetic stirrer, introduce the magnetic bead and heat the mixture gently with stirring. Add sodium potassium tartrate (30.0 g) in small lots. Make up the volume up to mark with distilled water.

20. **2,4-DNP in alcohol:** Suspend 2,4-DNP reagent (2.0 g) in methanol (60.0 mL), add carefully concentrated sulphuric acid (4.0 mL). The mixture gets warm and solid dissolves completely, if not, filter the undissolved substances. Store in an amber coloured bottle.

21. **2,4-DNP in water:** Suspend 2,4-DNP reagent (4.0 g) in a mixture of concentrated hydrochloric acid (35.0 mL) and water (65.0 mL). Warm on the water bath, if required. Cool and filter the solution. Store in an amber coloured bottle.

22. **Ethylenediamine tetra acetic acid solution (EDTA, 0.0015 M):** In a reagent bottle (100 mL), dissolve EDTA (0.0056 g) in a minimum amount of water and make up the volume up to mark with distilled water.

23. **Fehling solution**

    Fehling solution A: Dissolve copper sulphate pentahydrate (17.320 g) in distilled water (~ 100 mL) in standard flask (250 mL), add a few drops of 2.0 N sulphuric acid and make up the volume up to the mark. (Sulphuric acid prevents hydrolysis of copper sulphate!)

    Fehling solution B: Dissolve crystalline Rochelle salt (86.5 g) in minimum amount of hot water. Add sodium hydroxide solution (prepared by dissolving 30.0 g of sodium hydroxide in minimum amount of water). Mix the two solutions, cool and transfer into in a standard flask (250 mL). Make up the volume up to the mark with distilled water.

Keep both the solutions separately in tightly stoppered bottles and mix equal volumes before use.

24. **Ferric chloride solution:** In a clean standard flask (500 mL), dissolve ferric chloride hexahydrate (37.0 g) in minimum amount of water. Add concentrated hydrochloric acid (5.0 mL) and dilute up to mark with distilled water.

25. **Ferric chloride (Neutral) solution:** Take bench ferric chloride solution (2.0 mL) in a test tube and add dropwise sodium hydroxide solution till a slight excess of a brown coloured ferric hydroxide is formed. Filter the solution and use the filtrate.

26. **Ferroin reagent (1,10-Phenanthroline ferrous sulfate):** In a clean reagent bottle, dissolve 1,10-o-phenanthroline monohydrate (1.48 g) and $FeSO_4.7H_2O$ (0.70 g) in distilled water (100 mL).

27. **Ferrous sulfate reagent (5%):** Add ferrous ammonium sulfate crystals (5.0 g) to recently boiled, distilled water (100 mL). Add concentrated sulfuric acid (0.4 mL). Add an iron nail to retard air oxidation. Tightly stopper the reagent bottle and use immediately.

28. **Folin's reagent:** Take commercial sample (2.0 N) from the bottle, dilute 1:1 in a boiling tube, before use. (Keep away from sunlight. Use foil to cover the boiling tube).

29. **Iodine-potassium iodide reagent:** In a clean reagent bottle, dissolve iodine (10.0 g) and potassium iodide (20.0 g) in distilled water (100 mL).

30. **Jones reagent:** In a clean reagent bottle (100 mL), pour a suspension of chromic anhydride (25.0 g) in concentrated sulfuric acid (25.0 mL), slowly with stirring, into distilled water (75.0 mL). Cool the deep orange-red solution to room temperature before use.

31. **Lead acetate solution (0.5 N):** In a clean reagent bottle (100 mL), dissolve anhydrous lead acetate (16.2 g) in distilled water (100 mL) and store in reagent bottle.

32. **Lucas reagent:** In a clean standard flask (50.0 mL), dissolve anhydrous zinc chloride (13.6 g, 0.1 mol) in concentrated hydrochloric acid (10.0 mL, 0.1 mol), with cooling.

33. **Maltose solution (1 mg/mL):** In a clean and dry standard flask, dissolve maltose (0.250 g) in distilled water (250 mL) and store.

34. **Manganese sulphate solution:** In a clean standard flask (100 mL), dissolve manganese sulphate, $MnSO_4.H_2O$ (36.4 g), in water (100 mL). Shake and use.

35. **Methyl orange:** Dissolve methyl orange (1.0 g) in boiling water (1500 mL), filter, if required. Cool and store in reagent bottle.

36. **Molisch reagent:** Dissolve 1-naphthol (10.0 g) in 95% ethanol (100.0 mL) and store in a reagent bottle.

37. **Nickel Chloride and 2-Hydroxy-5-Nitrobenzaldehyde Reagent:** In a clean reagent bottle, add a solution of 2-hydroxy-5-nitrobenzaldehyde (0.5 g) dissolved in water (25.0 mL) to triethanolamine (15.0 mL). Then add nickel chloride hexahydrate (0.5 g) dissolved in water (10.0 mL) and bring the total volume of the solution to 100 mL with distilled water. If the triethanolamine contains ethanolamine, it may be necessary to add another 0.5 g of the aldehyde and remove the resulting precipitate by filtration.

38. **Ninhydrin reagent (0.25%):** In a clean reagent bottle (100 mL), dissolve ninhydrin (0.25 g) in rectified spirit (100 mL), shake and use.

39. **Phenolphthalein:** Dissolve phenolphthalein (1.0 g) in 50% aqueous ethanol (100 mL). Stopper the reagent bottle and use when required.

40. **Potassium bromate-bromide reagent (0.2N):** In a clean standard flask (1000 mL), dissolve potassium bromate (5.567 g) in water (700-800 mL) then add potassium bromide (75.0 g) to the same flask and finally make up to the mark with distilled water.

41. **Potassium Hydroxide (Alcoholic):** In a clean standard flask (100 mL), dissolve potassium hydroxide (3.0 g) in distilled water (3.0 mL). Make up the volume up to the mark with 95% ethanol. Stopper the flask and leave overnight. Filter, if required. This reagent is useful for carrying out ferrous hydroxide test for nitro compounds.

42. **Potassium Permanganate (1% solution):** Dissolve potassium permanganate (1.0g) in 100 ml of water in a reagent bottle and store.

43. **Schiff's reagent:** In a clean reagent bottle (250 mL), dissolve *p*-rosaniline hydrochloride (0.2 g) in warm water (50.0 mL), cool and saturate the solution by passing a stream of sulphur dioxide. Replace the stopper and allow the solution to stand for a few hours until the solution becomes pale yellow in colour. Dilute the solution with water (150 mL). Store in a tightly stoppered reagent bottle. Ensure that the solution is colourless.

Alternatively, take a clean and dry reagent bottle (200 mL), dissolve *p*-rosaniline hydrochloride (0.2 g) and concentrated hydrochloric acid (2.0 mL). Dilute the contents with distilled water (200 mL). Add sodium metabisulphite (2.0 g), stir and allow the solution to stand

and replace the stopper. The solution slowly becomes pale yellow/ colourless.

Note: If stopper is not tight, the solution gradually loses sulphur dioxide and regains pink colour.

44. **Seliwanoff's reagent:** Dissolve resorcinol (0.05 g) in concentrated hydrochloric acid (100.0 mL) and dilute very carefully with equal volumes of water.

45. **0.1M silver nitrate solution:** In a clean reagent bottle, dissolve silver nitrate (16.987 g) in water (1000 mL). Store in tightly stoppered amber coloured bottle.

46. **Sodium acetate (fused):** Take anhydrous sodium acetate (3.0 g) in a dry china dish and heat on a wire gauze. Stir continuously. The solid liquefies first, solidifies and then melts again. Allow the melted solid to cool (stir while cooling) and use immediately.

47. **1% Aqueous sodium chloride:** In a clean regent bottle (100 mL), dissolve sodium chloride (1.0 g) in acetate buffer (100 mL, pH-5.6). This reagent is useful for biochemical estimations.

48. **10% Aqueous sodium hydroxide:** In a clean reagent bottle (100 mL), dissolve sodium hydroxide (12-15 g) in water (100 mL). Shake and store.

49. **15% Sodium iodide in acetone:** Dissolve sodium iodide (15.0 g) in acetone (100 mL). The solution, colorless at first, becomes a pale lemon yellow. Keep the solution in a dark bottle and discard as soon as a definite red-brown color develops.

50. **Sodium nitroprusside solution (0.1%):** Dissolve sodium nitroprusside (0.1 g) in water (100 mL) taken in a reagent bottle. Use immediately, do not store.

51. **Starch solution (1%):** In a clean reagent bottle, dissolve starch (1.0 g) in boiling water (100 mL). Continue heating till the solution becomes almost clear. Cool and use.

52. **70% Sulphuric acid:** Add concentrated sulphuric acid (40.0 mL, AR) in small lots with stirring and cooling to water (30.0 mL) in a beaker.

53. **Tollen's reagent:** Prepare a solution of silver nitrate (30.0 g) in water (300 mL) and store in a reagent bottle. Whenever required, mix silver nitrate solution (1.0 mL) with 10% sodium hydroxide (1.0 mL) to obtain brown precipitate of silver oxide. Dilute the solution with aqueous ammonium hydroxide until the precipitate just dissolves. Add

a few additional drops of ammonium hydroxide to make the solution ammoniacal.

Note: The reagent decomposes on standing due to loss of ammonia. Therefore, it should be prepared fresh before use.

54. **Tris-EDTA buffer:** In a clean standard flask (100 mL), mix a solution of Tris (1.0 mL, 1 M), EDTA (0.0372 g) and dilute with distilled water (100 mL). Replace the stopper and store.

55. **Wij's solution:** Prepare a solution of iodine trichloride (7.9 g) in glacial acetic acid (100 mL) in a conical flask by warming on a water bath. In another flask take iodine (8.7 g) and dissolve in another 100 mL of warm glacial acetic acid. Mix the two solutions in a 1000 mL volumetric flask and make up the volume with glacial acetic acid. Store the resulting Wij's solution in a stoppered amber coloured bottle. (This solution is stable for one month!)

56. **Zinc chloride (Fused):** Take zinc chloride (2.0 g) in a clean and dry china dish and heat on a low flame using wire gauze. The compound first loses the absorbed water then melts and starts giving off white fumes. Remove from the burner, stir continuously until the compound solidifies. Use immediately.

APPENDIX III

Common Buffers Used in the Laboratory

The buffers described in this section are suitable for use either in enzymatic or histochemical studies. The accuracy of the tables is within ± 0.05 pH at 23°C. In most cases the pH values will not be off by more than ± 0.02 pH even at 37°C and at molarities slightly different from those given.

1. Citrate Buffer.

Stocks solutions

A: 0.1 M solution of citric acid (21.01 g in 1 L)

B: 0.1 M solution of sodium citrate (29.41 g $C_6H_5O_7Na_3.2H_2O$ in 1 L.

x mL of A + y mL of B, diluted to a total of 100 mL

x	y	pH
46.5	3.5	3.0
43.7	6.3	3.2
40.0	10.0	3.4
37.0	13.0	3.6
35.0	15.0	3.8
33.0	17.0	4.0
31.5	18.5	4.2
28.0	22.0	4.4
25.5	24.5	4.6
23.0	27.0	4.8
20.5	29.5	5.0
18.0	32.0	5.2
16.0	34.0	5.4
13.7	36.3	5.6

11.8	38.2	5.8
9.5	41.5	6.0
7.2	42.8	6.2

2. Acetate Buffer.

Stocks solutions

A: 0.2 M solution of acetic acid (11.55 g in 1 L)

B: 0.2 M solution of sodium acetate (16.4 g $C_2H_3O_2Na$ or 27.2 g of $C_2H_3O_2Na.3H_2O$ in 1 L).

x mL of A + y mL of B, diluted to a total of 100 mL

x	y	pH
46.3	3.7	3.6
44.0	6.0	3.8
41.0	9.0	4.0
36.8	13.2	4.2
30.5	19.5	4.4
25.5	24.5	4.6
20.0	30.0	4.8
14.8	35.2	5.0
10.5	39.5	5.2
8.8	41.2	5.4
4.8	45.2	5.6

3. Citrate-Phosphate Buffer.

Stocks solutions

A: 0.1 M solution of citric acid (19.21 g in 1 L)

B: 0.2 M solution of dibasic sodium phosphate (53.65 g of $Na_2HPO_4.7H_2O$ or 71.7 g of $Na_2HPO_4.12H_2O$ in 1 L).

x mL of A + y mL of B, diluted to a total of 100 mL

x	y	pH
44.6	5.4	2.6
42.2	7.8	2.8
39.8	10.2	3.0
37.7	12.3	3.2
35.9	14.1	3.4
33.9	16.1	3.6
32.3	17.7	3.8

30.7	19.3	4.0
29.4	20.6	4.2
27.8	22.2	4.4
26.7	23.3	4.6
25.2	24.8	4.8
24.3	25.7	5.0
23.3	26.7	5.2
22.2	27.8	5.4
21.0	29.0	5.6
19.7	30.3	5.8
17.9	32.1	6.0
16.9	33.1	6.2
15.4	34.6	6.4
13.6	36.4	6.6
9.1	40.9	6.8
6.5	43.6	7.0

4. Phosphate Buffer.

Stocks solutions

A: 0.2 M solution of monobasic sodium phosphate (27.8 g in 1 L)

B: 0.2 M solution of dibasic sodium phosphate (53.65 g of $Na_2HPO_4.7H_2O$ or 71.7 g of $Na_2HPO_4.12H_2O$ in 1 L).

x mL of A + y mL of B, diluted to a total of 200 mL

x	y	pH	x	y	pH
93.5	6.5	5.7	45.0	55.0	6.9
92.0	8.0	5.8	39.0	61.0	7.0
90.0	10.0	5.9	33.0	67.0	7.1
87.7	12.3	6.0	28.0	72.0	7.2
85.0	15.0	6.1	23.0	77.0	7.3
81.5	18.5	6.2	19.0	81.0	7.4
77.5	22.5	6.3	16.0	84.0	7.5
73.5	26.5	6.4	13.0	87.0	7.6
68.5	31.5	6.5	10.5	90.5	7.7
62.5	37.5	6.6	8.5	91.5	7.8
56.5	43.5	6.7	7.0	93.0	7.9
51.0	49.0	6.8	5.3	94.7	8.0

5. Tris(hydroxymethyl) aminomethane (Tris) Buffer.

Stocks solutions

A: 0.2 M solution of tris(hydroxymethyl) aminomethane (24.2 g in 1 L)

B: 0.2 M HCl

50 mL of A + x mL of B, diluted to a total of 200 mL

x	pH
5.0	9.0
8.1	8.8
12.2	8.6
16.5	8.4
21.9	8.2
26.8	8.0
32.5	7.8
38.4	7.6
41.4	7.4
44.2	7.2

Bibliography

1. *A Handbook of Organic Analysis* by H T Clarke, edition 2007.

2. *Practical Organic Chemistry* by F G Mann and B C Saunders, Edition 2009.

3. *Advanced Experimental Organic Chemistry* by V K Ahluwalia and Sunita Dhingra, Edition 2017

4. *Experimental Organic Chemistry* by Sonia Ratnani and Shriniwas Gurjar, Edition 2012.

5. *The Systematic Identification of Organic Compounds* by Ralph L. Shriner, Christine K. F. Hermann, Terence C. Morrill, David Y. Curtin, Reynold C. Fuson.

6. *Techniques in Organic Chemistry,* by Jerry R. Mohrig, Christina Noring Hammond, Paul F. Schatz, and Terence C. Morrill.

7. *Qualitative Organic Analysis: An Efficient, Safer, and Economical Approach to Preliminary Tests and Functional Group Analysis,* Sunita Dhingra* and Chetna Angrish, dx.doi.org/10.1021/ed1004454, *J. Chem. Educ.* 2011, 88, 649–651.

8. *Laboratoty Manual, Workshop on theory and practical course Biochemistry and Environment Chemistry,* B.Sc. (H) Chemistry, 2012, Department of Chemistry, University of Delhi.

9. *Green chemical processing in the teaching laboratory: a convenient liquid CO_2 extraction of natural products,* McKenzie, L. C., Thompson, J. E., Sullivan, R., & Hutchison, J. E., Green Chemistry, 2004, 6(8), 355. doi:10.1039/b405810k.

10. Furnis, B. S.; Hannaford, A. J.; Smith, P. W. G.; Vogel's Textbook on Practical Organic Chemistry, Fifth Edition, Longman, 1996.

Index